Wenqing Liu

Emulation of Narrowband Powerline Data Transmission Channels and Evaluation of PLC Systems

Forschungsberichte aus der Industriellen Informationstechnik
Band 7

Institut für Industrielle Informationstechnik
Karlsruher Institut für Technologie
Hrsg. Prof. Dr.-Ing. Fernando Puente León
       Prof. Dr.-Ing. habil. Klaus Dostert

Eine Übersicht über alle bisher in dieser Schriftenreihe erschienenen Bände finden Sie am Ende des Buchs.

Emulation of Narrowband Powerline Data Transmission Channels and Evaluation of PLC Systems

by
Wenqing Liu

Dissertation, Karlsruher Institut für Technologie (KIT)
Fakultät für Elektrotechnik und Informationstechnik
Tag der mündlichen Prüfung: 04. Juni 2013
Referenten: Prof. Dr.-Ing. habil. Klaus Dostert,
            Prof. Dr.-Ing. Martin Doppelbauer

Impressum

Karlsruher Institut für Technologie (KIT)
KIT Scientific Publishing
Straße am Forum 2
D-76131 Karlsruhe
www.ksp.kit.edu

KIT – Universität des Landes Baden-Württemberg und
nationales Forschungszentrum in der Helmholtz-Gemeinschaft

KIT Scientific Publishing 2013
Print on Demand

ISSN 2190-6629
ISBN 978-3-7315-0071-1

Emulation of Narrowband Powerline Data Transmission Channels and Evaluation of PLC Systems

Zur Erlangung des akademischen Grades eines

DOKTOR-INGENIEURS

von der Fakultät für
Elektrotechnik und Informationstechnik
des Karlsruher Instituts für Technologie (KIT)

genehmigte

DISSERTATION

von

Dipl.-Ing. Wenqing Liu
aus Hunan, China

Tag der mündlichen Prüfung: 04. Juni 2013
Hauptreferent: Prof. Dr.-Ing. habil. Klaus Dostert, KIT
Korreferent: Prof. Dr.-Ing. Martin Doppelbauer, KIT

Table of Contents

Acknowledgements

First of all, I sincerely thank my supervisor Prof. Dr.-Ing. habil. Klaus Dostert for his valuable advice, continuous support and for being always available for open discussions. His personal guidance, constructive comments and broaden knowledge have helped me greatly throughout my research and during the writing of this thesis.

I am also extremely indebted to Prof. Dr.-Ing. Martin Doppelbauer for accepting to be my second supervisor, for his interest in my work and for his valuable advice to improve this thesis.

I am deeply grateful to Prof. Dr.-Ing. Fernando Puente and all my colleagues at Institute of Industrial Information Technology (IIIT) for their encouragement, for their help and for providing me with a pleasant and friendly working atmosphere. Further, I would like to acknowledge the personnel and the technique staff at IIIT for their great support in administrative matters. It would not have been possible to carry out the research without their unstinting support.

I also owe much gratitude to Dipl.-Ing. Martin Sigle, Dipl.-Ing. Oliver Opalko and Dr. Jinshi Li, for their valuable suggestions for improving the thesis.

I would like to thank all of my students who have contributed to my research work in the form of study projects, graduate theses and research assistance. I am also grateful to all my teaching assistants for their important support in leading the course "microcontrollers and digital signal processors".

Last but not least, I would like to express my greatest thanks to my family members, especially my wife Qian. Their permanent support, great encouragement, infinite patience and meticulous care played a very important role in the completion of this thesis.

Abstract

Reliable communication plays a very important role in smart grid applications. Narrowband powerline communication (NB-PLC) technologies in the frequency range up to 500 kHz are becoming more and more popular and many commercial products are available. However, customers and system developers are challenged by performance evaluation of the NB-PLC systems. The difficulties are mainly raised by the complexity of channel characteristics. This thesis proposes advanced emulation of the physical layer behavior of NB-PLC channels and the application of a channel emulator for the evaluation of NB-PLC systems. The emulator is able to reproduce PLC-related channel conditions, such as bidirectional transfer functions, sophisticated noise scenarios and the corresponding time varying characteristics. Based on the channel emulator, a testbed is developed to provide the real-world mains environment. Any systems can be plugged and tested, just as they are applied in a real-world channel. In addition, test procedures and reference channels are proposed to improve efficiency and accuracy in the system evaluation and classification. To illustrate the power of the physical layer channel emulation, several case studies are also presented. This thesis shows that the channel characteristics can be reproduced under laboratorial conditions any time with the help of the channel emulator. It is very convenient to apply the emulator-based testbed to compare PLC systems that are based on different technologies. The channel emulator opens new ways toward flexible, reliable and technology-independent performance assessment of PLC modems.

Zusammenfassung

Bei der Entwicklung intelligenter Stromnetze, sogenannter Smart-Grids, spielt eine zuverlässige Kommunikation eine entscheidende Rolle. Die Technologien schmalbandiger Datenübertragung über Energieversorgungsleitungen im Frequenzbereich bis 500 kHz, in Englisch auch als „narrowband powerline communication (NB-PLC)" bezeichnet, werden immer beliebter für Smart-Metering- und Steuerungsanwendungen. Dafür ist auch eine zunehmende Anzahl von NB-PLC-Lösungen verfügbar. Allerdings stehen über das Verhalten der Kommunikationssysteme in realen Datenübertragungskanälen nur wenige Informationen zur Verfügung. Die Schwierigkeiten bei der Evaluierung der Leistungsfähigkeit des jeweiligen Systems liegen darin, dass einerseits die Eigenschaften des PLC-Datenübertragungskanals zeitvariant und von der jeweiligen Netztopologie abhängig sind, und anderseits noch keine Referenzkanäle zur Untersuchung von NB-PLC-Systemen verfügbar sind. Das Ziel der vorliegenden Arbeit besteht darin, ein echtzeitfähiges Evaluierungssystem zur Untersuchung von NB-PLC-Systemen zu entwickeln. Ein wesentlicher Bestandteil dieses Systems ist ein sogenannter Kanalemulator, welcher die für NB-PLC relevanten Kanaleigenschaften im Frequenzbereich bis 500 kHz nachbildet. Die zu emulierenden Kanaleigenschaften bestehen hauptsächlich aus der Kanalübertragungsfunktion und dem komplexen Störszenario. Die Übertragungsfunktion beinhaltet sowohl Amplituden- als auch Phasengang. Der Emulator ist für eine bidirektionale Datenübertragung ausgelegt, d.h. es wird Übertragungsfunktion und Störszenario sowohl für den Hin- als auch für den Rückkanal nachgebildet. Um den Kanal möglichst genau zu reproduzieren, muss die Zeitvarianz der Kanaleigenschaften berücksichtigt werden. Zusätzlich zum Emulator werden weitere Systemkomponenten entwickelt bzw. optimiert, um das Evaluierungssystem vom realen Versorgungsnetz zu isolieren. Darüber hinaus werden Testverfahren und Referenzkanäle vorgeschlagen, um die Effizienz und Genauigkeit der Testergebnisse zu verbessern. Am Ende stehen der Aufbau des Evaluierungssystems für die Untersuchung verschiedener NB-PLC-Systeme und mehrere Fallstudien zur Veranschaulichung des Einsatzes und der Fähigkeiten des Evaluierungssystems. Die Arbeit zeigt, dass das Evaluierungssystem es ermöglicht, PLC-Übertragungskanäle jederzeit in einer Laborumgebung zu emulieren und unterschiedliche NB-PLC-Lösungen unter realistischen und reproduzierbaren Bedingungen zu vergleichen.

1 Introduction

1.1 Motivation

Reliable communication plays a key role in smart grid applications like advanced meter reading (AMR), load control and remote diagnostics. Narrowband powerline communication (NB-PLC) technologies in the frequency range up to 500 kHz are becoming more and more popular. Generally, narrowband refers to the frequency range 3-148.5 kHz in Europe, defined by Comité Européen de Normalisation Électrotechnique (CENELEC), 10-490 kHz in the USA, 10-450 kHz in Japan and 3-500 kHz in China [GAL11]. With the advent and popularity of NB-PLC technologies, many commercial products are available.

Generally speaking, different technologies are suitable for different transmission channels. Even systems that are based on the same standard can perform differently if they are provided by different manufacturers. However, there is very limited information on the performance of the NB-PLC systems under representative real-world channel conditions. This limitation is mainly caused by the complexity of channel characteristics. Unlike dedicated data transmission links, the NB-PLC channels are characterized by high dynamics and high attenuation in the amplitude response [BAU06]. Frequency-selective fading can occur and the fading characteristics depend on load situations. Due to the time-variant nature of the electrical loads, the transfer functions and the noise scenarios also change with time [CAN06]. In addition, the channel behavior exhibits geographical dependency and a wide range of diversity. Unfortunately, there are no commonly adopted reference channels in this frequency range for the purpose of system test and verification. As a result, system designers face the difficulty of verifying their products and benchmarking systems under representative and fair conditions.

Besides reference channels, there are also no satisfying test platforms available. Many system designers and customers are still working with traditional methodologies and struggling for improvements in flexibility, efficiency and accuracy. Fig. 1.1 shows three traditional approaches to test communication systems. The

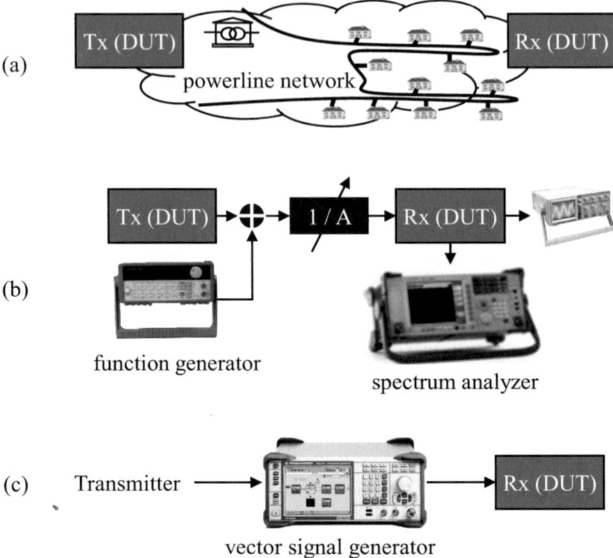

Fig. 1.1 Traditional approaches to test communication systems. (a) connecting the modems to mains networks directly. (b) test platform composed of switchable attenuator and standard laboratory equipment. (c) vector signal generator-based test environment.

most realistic way is to connect the modems to the mains network directly, as shown in plot (a). However, this suffers from bad reproducibility caused by the time-varying and complicated channel conditions. The classification of modems is therefore neither reliable nor convincing. Nowadays, more and more tests are made under laboratory conditions. The commonest and simplest method is to connect the transmitter and the receiver using a switchable attenuator, as shown in plot (b). A function generator is utilized to generate artificial noise. The transmitted signal and the noise can be monitored using a spectrum analyzer or an oscilloscope [JEN07]. However, this test setup cannot create complicated and realistic channel conditions. In particular, the attenuator has a flat attenuation over the frequency range of interest, thus is not able to produce the aforementioned frequency dependency. The shortcoming could be partially overcome by inserting L-C-R resonance circuits between the devices under test (DUT). Nevertheless, this makeshift is of low flexibility, low accuracy and high complexity. Vector signal generators are very popular in mobile phone and wireless tests. As illustrated in plot (c), one can build a transmitter by such a device, and add noise and attenuation to the transmitted signal, so that the signal seems to be distorted similarly as in a real channel. The method is accurate and flexible. However, exact details of the DUT's parameters are needed. In order to test several different devices, an individual transmitter must be provided for each one. The solution is therefore technology-dependent.

Traditional equipment and approaches make the test of NB-PLC systems quite challenging. A workaround, as shown in Fig. 1.2, is to build an isolated powerline testbed and to mimic the real-world PLC-related behavior, using a stand-alone device which will be called channel emulator in the following. Channel characteristics can be reproduced under laboratory conditions any time. Since only the channel behavior is emulated, it is not necessary to acquire detailed information about the DUTs. Any PLC system can be connected to the testbed for performance evaluation, and the system evaluation is highly flexible and efficient. The channel emulator may indeed open a new path to flexible, reliable and technology-independent performance evaluation of PLC modems.

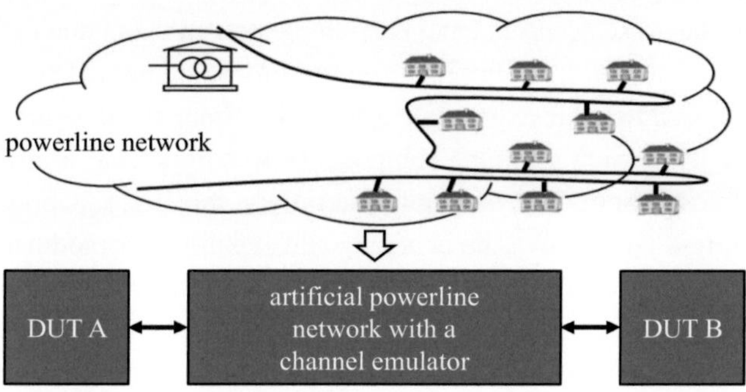

Fig. 1.2 Test of PLC systems using emulated channels.

1.2 State of the Art

The idea of emulating PLC channels using a stand-alone device is not new. Some concepts have already been presented for broadband PLC (BB-PLC) in the last few years. For example, a universal high-speed channel emulation system has been introduced in [GOE02]. This system is based on a digital signal processor (DSP) and a field programmable gate array (FPGA). Algorithms have been developed and optimized to reduce complexity and improve real-time performance.

Both the channel transfer functions and typical noise scenarios have been emulated in an unidirectional prototype. Later, this prototype has been extended to a bidirectional channel emulator. Full duplex modem operations can be evaluated, and asymmetric channel behavior can be applied for modem tests. In addition, nine reference channels have been proposed for the access domain within the Open PLC European Research Alliance (OPERA) project of the European Union [BAB05] [FER10]. A channel emulator for indoor channels in the frequency range up to 30 MHz has been reported in [CAN08]. It is based on a comprehensive channel model and can emulate additional features including time-varying channel attenuation and cyclo-stationary noise scenarios. A flexible FPGA-based emulator has been described in [WEL11]. It can emulate complex scenarios such as multiple-input and multiple-output (MIMO) transmission, neighborhood network, hidden node effect and very high speed digital subscriber line (VDSL) co-existence scenarios.

For the NB-PLC channels, an emulator has been realized for indoor channels in the frequency range between 95 and 148.5 kHz [BAU052]. The channel transfer function and the noise scenario have been implemented using analogue and digital circuits respectively. In addition, a new test procedure named "Orphelec Test" has been proposed to achieve accurate test results. Since the transfer function has been realized by passive analogue components, it suffers from limited flexibility and it is almost impossible to emulate arbitrary transfer functions and time-varying features. To improve the accuracy and flexibility, a modular PLC channel emulator has been developed in [LIU11], with which more sophisticated channel characteristics can be reproduced. However, the emulated channel transfer function exhibits relatively low frequency resolution. Furthermore, only one direction can be emulated.

1.3 Goal and Challenges

The goal of this work is to develop an advanced channel emulator for NB-PLC on which a flexible testbed can be implemented for the evaluation of PHY-layer performance of NB-PLC systems. The emulator shall be capable of providing transfer functions for bidirectional data transmission. It shall be able to generate sophisticated noise scenarios and reproduce the corresponding time varying characteristics. In addition, test procedures and reference channels shall be developed to improve test efficiency and accuracy. In order to reach the goal, several challenges are to overcome:

Complex-Valued Channel Transfer Function

Investigations of transfer functions of NB-PLC channels, especially in the access domain, are usually limited to magnitude response. There is very limited information about phase response and group delay. The channel is often considered to exhibit linear or random phase responses. However, the channel phase may not be a linear function of the frequency, due to the complexity of the channel conditions. The phase response may even change with time, due to the influence of the connected electrical appliances. Periodic phase changes that are synchronized with the mains cycle have been observed in [SUG09]. Therefore, it is insufficient to treat NB-PLC channels as linear or random phase systems. It is necessary to handle the complex-valued channel transfer functions for an accurate emulation.

Definition of Reference Channels

Currently, there are no accurate reference channels for NB-PLC. A reference channel is a parameter set which describes a realization of the representative channel characteristics. Well-defined reference channels provide the opportunity to evaluate communication systems comprehensively and effectively. Some typical European indoor channels have been investigated in the frequency range of interest in [BAU05], and numerous reference channels have been proposed. Nevertheless, the reference transfer functions are limited solely to magnitude responses. Furthermore, sophisticated time-varying behavior, such as the periodic variations within each mains cycle is not treated. Therefore, the existing reference channels must be extended accordingly.

Emulating Real-World Channels

Besides the reference channels, it is sometimes desirable to emulate specific channels in which a NB-PLC system will be applied. It is important to measure those channels using appropriate methodologies, so that the most relevant features can be obtained. Proper models must then be applied to the measured data to obtain parameters for the channel emulation. Guidelines are expected for the whole process, so that any real-world channel can be reproduced as accurately as possible.

Bidirectional Channel Emulation

Bidirectional channel emulators are using directional couplers or hybrids to separate incident and reflected signals [WEL11]. A prerequisite for this approach to work is an accurate matching of the input impedance of the connected DUTs. The impedance is mainly determined by the analog components of the DUT, such as the analog front-end and the coupling circuits. Unfortunately the design and realization of these components are not standardized for NB-PLC systems. Different systems can have quite different input and output characteristics. Even systems provided by the same manufacturer can have diversity due to component tolerances. As a result, great efforts have to be made for the impedance matching for individual DUTs. The complexity becomes even higher if the DUTs are based on multicarrier technologies, because the impedance matching has to be achieved at every single subcarrier frequency. Therefore, an alternative approach is necessary for higher flexibility.

"Pure" Mains Voltage for Testbed

Performance evaluation must be carried out in presence of the mains voltage. Since the real-world mains network exhibits unpredictable behavior when considered as a data transmission channel, it is very important to prevent the testbed from being disturbed by unwanted effects and at the same time to provide the testbed with a "pure" mains voltage. It is a challenging task to "isolate" the testbed from the mains network and to suppress the various high-level noises at low frequencies.

1.4 Structure of the Thesis

The thesis is organized as follows: chapter 2 discusses two aspects of NB-PLC: the channel properties and the communication systems. Typical characteristics of physical channels, methodologies for channel modeling and the resulting models are reviewed. Meanwhile, fundamentals of different NB-PLC systems are studied. In chapter 3, a universal emulator-based testbed is proposed for the performance evaluation of communication systems. Chapter 4 presents acquisition and emulation of complex-valued channel transfer functions. Chapter 5 discusses extraction of patterns from measured noise and synthesis of the noise patterns using the channel emulator. In chapter 6, case studies are presented to illustrate the capability of the emulator and the testbed. In addition, accurate reference channels are proposed. Finally, chapter 7 concludes this thesis and raises discussions for future work.

2 Narrowband-PLC

This chapter discusses two aspects of NB-PLC: typical channel properties and communication systems making use of the channel. For the channel, typical networks will be studied. Existing methodologies for channel characterization and modeling will be examined. After that, an overview of typical NB-PLC communication systems will be given. Important system components that are customized for PLC applications will also be analyzed. This chapter is intended to provide necessary information from a communication system's point of view, so that the characterization and emulation can focus on the channel characteristics that are relevant for communications.

2.1 Channel Characterization and Modeling

An accurate characterization of the channel is a prerequisite for the successful design of any communication system. Fig. 2.1 illustrates the relation between the

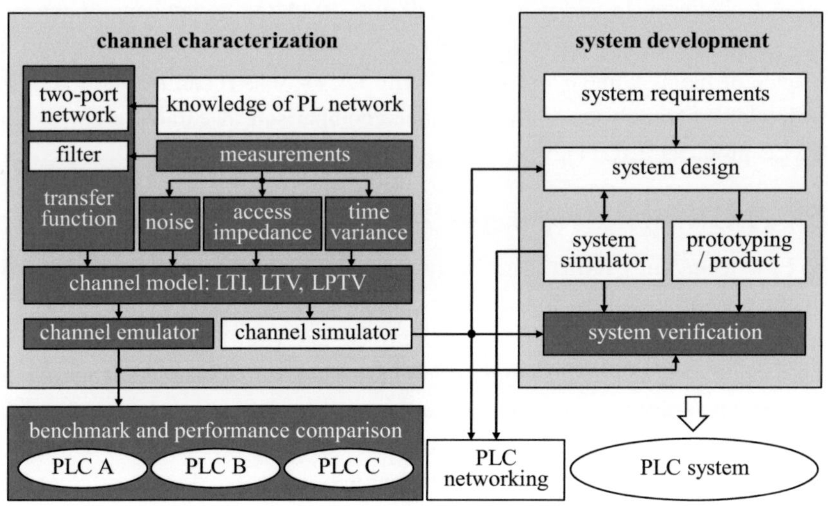

Fig. 2.1 Relation between channel characterization and system development.

exploration of powerline channels and the development of PLC systems. Typical activities for channel characterization include channel measurement, network topologies analysis, channel modeling, simulation and emulation. The PHY-layer behavior of a powerline channel can be described by its transfer function, noise scenario, access impedance and the time-variance. Two fundamental methodologies have been developed to model the transfer function: the statistical approach and the deterministic approach. The first is based on preliminary measurements and statistical analysis methodologies. The channel influence on the transmitted signal can be modeled using a filter. The latter develops a closed-form expression under consideration of detailed network knowledge, such as network topology, cable material, connected appliances, etc. Both approaches have their advantages and disadvantages, as discussed in great depth, e.g. in [FER10]. Due to the time-varying nature of the real-world channel, the model variants extend from a simplified linear time-invariant (LTI) over linear time-variant (LTV) and linear periodically time-variant (LPTV) descriptions. The simulator and the emulator can be considered as instances of a channel model. Most simulators are based either on closed-form expressions or on simulation software for circuit design, while the emulator is a stand-alone hardware implementation. Due to their different natures, both are applied differently. Together with a system simulator, the channel simulator can be used for system design activities, for example the evaluation of modulation techniques. In addition, the flexibility of adding components makes the study of complex PLC networking schemes possible. The emulator can work in presence of mains voltage and interact with DUTs, thus it can be applied in the verification of prototypes and performance comparison of commercially available PLC products [LIU11].

2.1.1 PLC Network Topologies

The PLC technology utilizes the existing wiring infrastructure directly. As a result, the network topology for communications depends largely on the mains wiring topology. An analysis of the medium voltage (MV) to low voltage (LV) mains topologies will help to understand the PLC channel characteristics. Mains topologies may differ from country to country. Even within a country, electricity may be supplied differently in rural and urban areas. Residential, industrial and business areas may also have different electricity networks due to different load situations. Despite these diversities, the mains topologies, and thus the PLC network topologies as well, can be classified into two groups: outdoor and indoor channels. The outdoor channel covers interconnects between transformer stations and house connections; it is also called the access domain. The indoor channel, as the name suggests, includes all connections inside buildings and houses.

2.1.1.1 Access Domain

A typical European MV/LV electricity topology is shown in Fig. 2.2. The 50 Hz/ 230 V 3-phase system is deployed in the LV grid. The households supplied by a transformer station are organized as a so-called supply cell which consists of up to 350 households. Each cell can include as many as 10 branches and each branch serves around 30 households. Depending on the load intensity, multiple transformers can be used within a substation. A branch may reach a maximum length of 1 km, e.g. in case that one family houses dominate the load scenario. Subgroups of several households are always connected with relatively short cables. A large number of parallel connections of appliances within each house can lead to low access impedance at the point where the powerline enters the house. The number of households in a rural network may be reduced [DOS01].

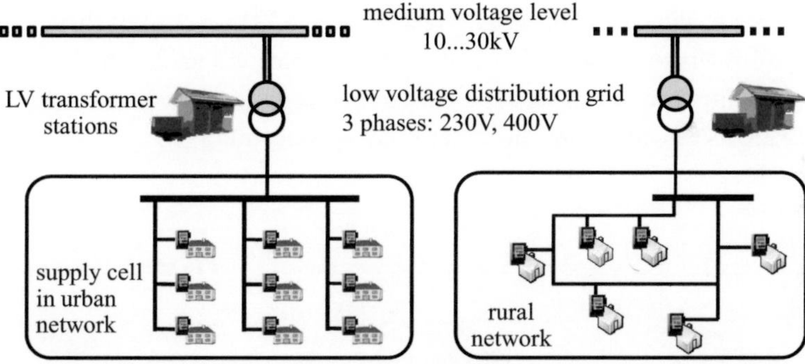

Fig. 2.2 Electricity supply topology in Europe in the access domain.

In the USA and some Asian countries, such as Japan, the 60 Hz/120V system is deployed in LV mains. In comparison with European networks, there are an increased number of distribution transformers. Furthermore, each distribution transformer has much fewer households to supply, in particular in rural areas. The cables connecting the transformer to the households exhibit around 100 m in length [FER10]. Due to the distinction in the wiring topologies, the PLC networks are constructed differently in different countries. Take the metering application for example. Fig. 2.3 shows a typical PLC network in Europe and China. A NB-PLC modem is integrated in each smart meter at the customer's premises (households). Each modem collects the meter information and sends it to a data concentrator installed at the LV side of the distribution transformer. The data concentrator is connected to the utility company through a telecommunication backbone.

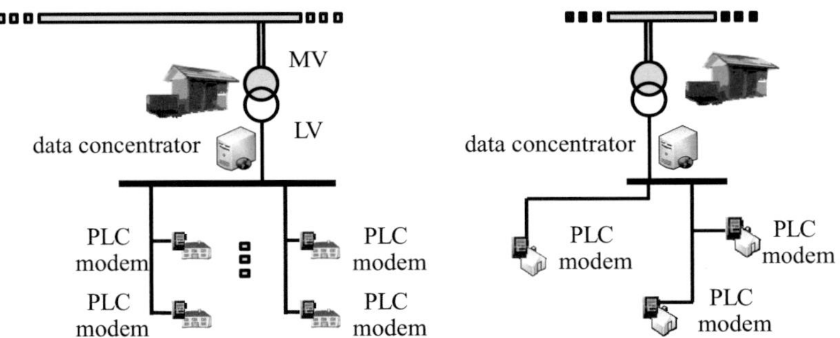

Fig. 2.3 PLC network construction in Europe and some Asian countries, such as China, for metering applications.

In the USA, the data concentrator resides on the MV instead of the LV side of the transformers to cut overall system cost, as shown in Fig. 2.4. In this situation, the communication signal must be able to penetrate the transformers so that the PLC modems can communicate with the data concentrator.

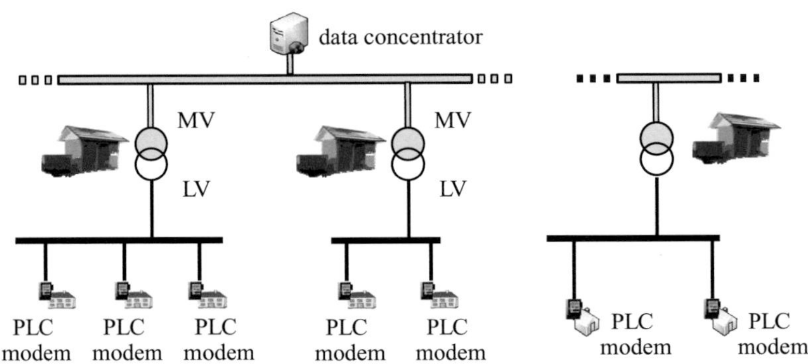

Fig. 2.4 PLC network construction in the USA for metering applications. The data concentrator resides on the MV level. Communication signals have to cross the transformers so that data links can be created between the data concentrator and the modems.

2.1.1.2 Indoor Channel

Although the indoor channel is mainly "occupied" by BB-PLC systems, there are still NB-PLC applications for home-automation [SUG08][BAU06]. It is worth examining the indoor electricity topologies as well. A household can be supplied by a single phase, such as in the USA and Japan [SUG08], or multiple phases such as in Europe. The wiring topologies inside homes and small offices have tree, star or even ring configurations.

In single-phase systems, power cables can have three or four conductors. Fig. 2.5 shows a wiring diagram of a single-phase system with three conductors, including two live conductors and a neutral wire. Each of the hot wires carries a 60 Hz alternating current (AC) mains voltage with 120 V_{rms}. The voltage waveforms are out of phase and symmetric with respect to the ground potential. High power appliances such as air conditioners and dryers can be connected to both live conductors and thus can obtain the doubled nominal voltage. A main breaker or main fuse can break the electrical connection and isolate the household from the power supply network. The main breaker is followed by an electricity meter which measures the power consumption. A service panel behind the electricity meter feeds branches. There are circuit breakers for each branch. Each branch can be connected to one of the two live conductors and the neutral, or to both live conductors.

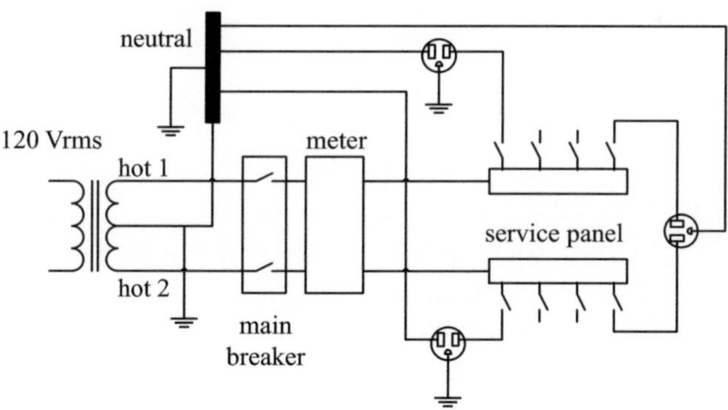

Fig. 2.5 Wiring diagram for a single-phase system.

Considering the possibility that NB-PLC systems can be connected to any receptacles, there are three kinds of communication links: links in the same branch, links on the same phase but consisting of different branches, and cross-phase links. In the three-phase indoor wiring topologies, the rooms of a household may be on different phases. Fig. 2.6 illustrates a simplified wiring diagram containing both ring and tree topologies. The ring circuit is deployed widely in the UK. Three conductors, one for live, one for neutral and the last one for earth, starting and ending at the service panel, form a closed ring topology. Sockets and appliances connected to the ring can obtain power via two paths, in this way, the network still works, even if the cables are broken somewhere, and thus consequences of faults are reduced. In the tree configuration, the service panel is considered as the root of the tree. Branches and splits are made to extend the coverage. The tree topology provides only one path between any set of connected nodes. The node at the end of a branch has the largest electrical distance to the service panel.

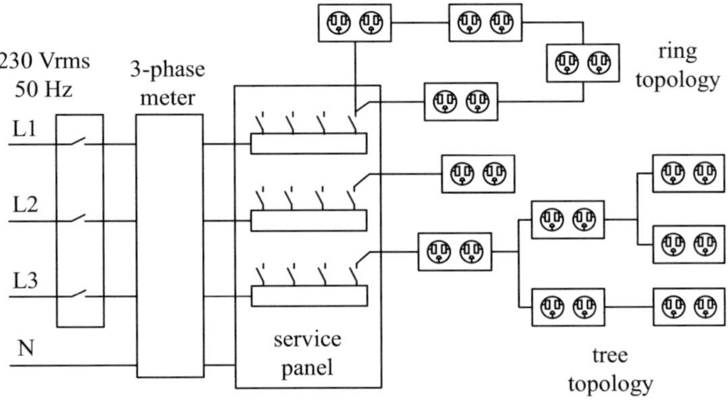

Fig. 2.6 Wiring diagram for a three-phase system.

2.1.2 Structural Model

Based on the aforementioned network topologies, a loaded transmission line model can be derived by considering wiring cables, electrical appliances and PLC systems. As shown in Fig. 2.7, a PLC transmitter is simplified using a signal source s_T and its equivalent output impedance Z_T. A receiver s_R is represented by its equivalent input impedance Z_R. There is a direct path between the transmitter and the receiver. In addition, a bridged tap interprets a branching connection along the direct path. Power cables connecting any two devices are modeled by two port networks (2PNs). A device can be an electrical appliance or a PLC modem. An electrical appliance is modeled by a noise source n_i with its series impedance Z_i. Each 2PN can be represented by an ABCD matrix which describes the relationship between current and voltage at the in- and output ports. The ABCD coefficients are complex functions of frequency and determined mainly by the cable characteristics. The 2PN and cascaded 2PNs can be used to model both two-conductor and multi-conductor transmission lines [GAL06]. In addition, the transmitter and the receiver must not be on the same phase. The ABCD coefficients of the 2PNs are assumed to be time-invariant since the wiring topology is usually not changed. On the contrary, the series impedance and the noise pattern of the appliance model show both long- and short-term variations which will be discussed later. A similar structure has been reported in [SAN07] for simulating PLC channels for broadband applications. Nevertheless, it is limited to the two-conductor model. Furthermore, the feeding line connecting an appliance to the mains is also considered. In the NB frequency range, the feeding line is very short in comparison with the wavelength of the transmitted signal. Therefore the feeding line is ignored in the proposed model and the appliances are considered to be connected to the distribution line directly.

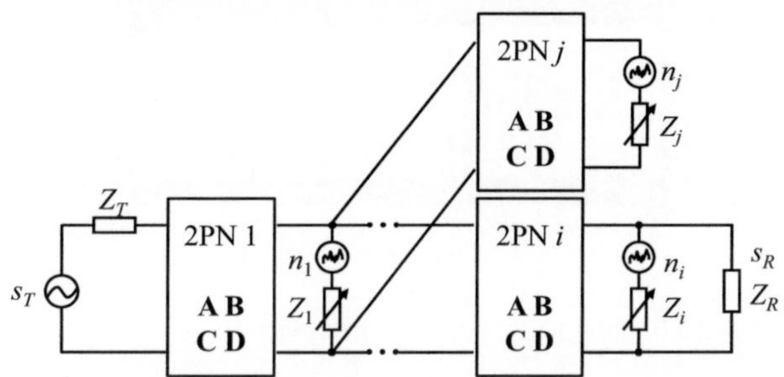

Fig. 2.7 Structural model of powerline networks.

It is extremely difficult to develop such an accurate structural model for an arbitrary powerline network. The difficulties are raised by many factors. It is almost impossible to obtain an accurate wiring topology with known cable parameters. The noise generated by the appliances and the impedance cannot be measured individually without the influences of overall cable characteristics and other appliances. Nevertheless, it gives valuable insights in the transfer characteristics of a PLC network and the origin of noise, and helps toward a better understanding of the time-varying nature of the channel behavior.

2.1.3 Behavioral Model

An expanded linear time-variant model is deployed to describe the behavior of the channel between a transmitter and a receiver. As shown in Fig. 2.8, it consists of access impedance, a linear filter and an additive noise source. t and f indicate the time and the frequency domains respectively. The time-varying access impedance $\underline{Z}_A(t, f)$ is the equivalent impedance seen into a power outlet. Together with the equivalent time-invariant output impedance $\underline{Z}_T(f)$ of the transmitter, it determines the signal level that can be injected into the powerline network. Let $x(t)$ and $s_A(t)$ denote the original and the injected signals respectively. $X(f)$ and $S_A(t, f)$ are their spectral components respectively.

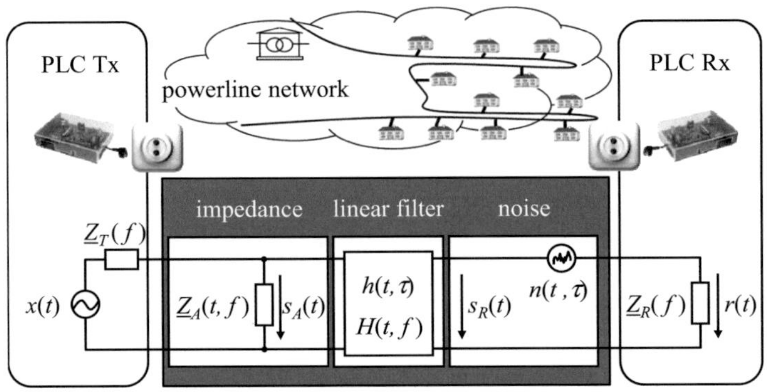

Fig. 2.8 Equivalent electrical channel model.

The magnitude ratio of $s_A(t)$ to $x(t)$ can be obtained by

$$H_A(t, f) = \frac{S_A(t, f)}{X(f)} = \frac{\underline{Z}_A(t, f)}{\underline{Z}_A(t, f) + \underline{Z}_T(f)}, \tag{2.1}$$

$H_A(t, f)$ is usually frequency-selective. Its magnitude is usually smaller than 1, therefore it is called coupling loss and is mainly caused by an impedance mismatching [HOO98]. Many efforts have been made to compensate the coupling loss. For example adaptive algorithms and special coupling circuitry have been developed to match the impedance during the transmission [CHO07]. The received signal $s_R(t)$ is usually the attenuated and distorted counterpart of $s_A(t)$. For a linear system, the relation between $s_R(t)$ and $s_A(t)$ can be described by a transfer function $H(t, f)$ in the frequency domain or by an impulse response $h(t,\tau)$ in the time domain. Generally speaking, $h(t,\tau)$ gives the response at time t caused by an impulse excitation applied at time t-τ. t and τ are also called the time of the observation instant and the time of excitation respectively [CLA82]. $s_R(t)$ is related to $s_A(t)$ by

$$s_R(t) = \int_{-\infty}^{\infty} h(t,\tau) \cdot s_A(t - \tau) d\tau. \tag{2.2}$$

$h(t,\tau)$ and $H(t, f)$ fulfill the relation

$$H(t, f) = \int_{-\infty}^{\infty} h(t,\tau) \cdot e^{-j2\pi f\tau} d\tau. \tag{2.3}$$

The additive noise $n(t,\tau)$ is a collection of all interference seen at the receiver. From the signal processing's point of view, the received signal $r(t)$ is obtained by

$$r(t) = x(t) * h_A(t,\tau) * h(t,\tau) + n(t,\tau) \tag{2.4}$$

in the time domain, or

$$R(f) = X(t,f) \cdot H_A(t,f) \cdot H(t,f) + N(t,f) \tag{2.5}$$

in the frequency domain.

2.1.4 Time Variations

The time varying features of powerline channels can be categorized in long- and short-term variations. The long-term variation is caused by electrical changes in powerline networks, e.g. switching on or off of electric appliances, or any topological changes due to load adjustment, management of distributed energy generation, and so on. Most of the changes happen randomly. Transitions from one state to the other are supposed to be fast and instantaneous. Within each state the channel topology is considered to be time-invariant. The channel coherence time, within which the channel's impulse response doesn't change, is around minutes or even hours and is multiple orders larger than the effective length of an individual impulse response [SAN07] [SIG11]. Therefore the long-term variation is also considered as a slow variation from the transition rate's point of view. This random process can be modeled by Markov chains which will be discussed in chapter 5.

The short-term variation is mainly caused by connected appliances during their operation. Their electrical parameters, such as the equivalent impedance which they present to the mains and the noise they generate, depend on the amplitude of mains voltage. This kind of varying characteristics can lead to periodic signal fading with a repetition rate of 100 Hz in 50 Hz or 120 Hz in 60 Hz systems [CHA86]. This periodicity in the transfer function can be described by a LPTV model, which will be introduced shortly. Further studies have shown that the variation is slow compared to the switching of transmitted signals [CAN06] [SIG11]. Therefore, the continuous time variation can be discretized by a series of states, in each of which the channel characteristics are considered to be invariant. Therefore, the channel responses can be approximated by a sequence of LTI responses.

2.1.5 Linear Periodically Time-Variant (LPTV) Model

[CAN06] and [SAN07] consider the short-term variation to have its origin in the non-linear nature of electrical devices. The non-linear behavior can be described by the Volterra series which is capable of relating the system output to its current as well as to previous inputs. Let $x(t)$ and $r(t)$ denote the input and output respectively. $r(t)$ can be expressed by

$$r(t) = h_0 + \sum_{n=1}^{\infty} \int \cdots \left[\int \left[\int h_n(\tau_1, \tau_2, ..., \tau_n) x(t - \tau_1) d\tau_1 \right] x(t - \tau_2) d\tau_2 \right] \cdots x(t - \tau_n) d\tau_n,$$

(2.6)

where $h_n(t_1, t_2, ... t_n)$ is a Volterra kernel which represents the nth order impulse response. Mathematically, it is actually a collection of an infinite sum of multidimensional convolutional integrals. Due to its memory effect, the behavior of some devices which can store electrical energy, for example capacitors and inductors, can be modeled with it. In PLC channels the transmitted signal is coupled to the mains voltage additively. At the receiver, the signal will be extracted using a bandpass filter, and the mains voltage will be strongly attenuated. Therefore the NL system can be simplified to

$$h(t,\tau) = h_1(\tau) + hm_1(t,\tau) + hm_2(t,\tau) + ..., \tag{2.7}$$

with

$$hm_1(t,\tau) = \int_{-\infty}^{\infty} h_2(\tau,\tau_1)m(t-\tau_1)d\tau_1 \tag{2.8}$$

and

$$hm_2(t,\tau) = \int_{-\infty}^{\infty}\left[\int_{-\infty}^{\infty} h_3(\tau,\tau_1,\tau_2)m(t-\tau_1)d\tau_1\right]m(t-\tau_2)d\tau_2. \tag{2.9}$$

$h_2(\tau, \tau_1)$ and $h_3(\tau, \tau_1, \tau_2)$ in (2.8) and (2.9) are the 2nd and the 3rd Volterra kernels respectively. Obviously the terms hm_1, hm_2 and higher orders give rise to a time-varying effect. All time behavior is associated with the mains voltage $m(t)$. A channel is a LPTV system when its impulse response changes repeatedly with a period of T_0.

$$h(t,\tau) = h(t,\tau - nT_0). \tag{2.10}$$

This can be true for PLC channels since $m(t)$ is a periodic function of time. This periodicity also appears in the frequency domain:

$$H(f,\tau) = H(f,\tau - nT_0). \tag{2.11}$$

2.1.6 Mains Voltage and Frequency

The waveform of the mains voltage $m(t)$ can be approximated at its fundamental frequency by

$$m(t) = (A_m + \Delta A_m) \cdot \sin\left[2\pi \cdot (f_m + \Delta f_m) \cdot t\right], \tag{2.12}$$

where f_m and A_m denote the nominal frequency and voltage respectively. Δf_m and ΔA_m refer to the deviations from f_m and A_m respectively. The nominal system parameters are standardized differently and they vary from region to region. Europe, Africa, most parts of Asia and South America use 230 V_{rms} allowing a tolerance of $\pm 6\%$. While in the USA, Japan and some parts of South America, a voltage between 100 and 127 V is used. With respect to the mains frequency, most regions and countries utilize 50 Hz, except e.g. North America and some parts of South America, where 60 Hz is common. Δf_m is a non-zero value, it depends on the relation between load and generation conditions. Regulation of daily average frequency is made to keep Δf_m within a given tolerance range, so that the flow of power from multiple generators through the network is under control. The maximum allowable deviation is standardized. For example, DIN EN 50160 averages f_m over 10 seconds. The resulting Δf_m must be smaller than 1% for synchronous mains networks during 99.5% of a year. This results in a maximum deviation of 0.5 Hz. Investigations of Δf_m have been made in [KIS09]. Zero-crossing jitter errors - defined as the deviations of the mains period from the nominal 20 ms - have been measured in Germany and Switzerland. It has been observed that 99.91 % of the jitter errors fall into an interval of ± 60 µs, and they exhibit almost a normal distribution within this range.

2.2 Typical PLC Systems

Different PLC systems may have different implementations. However, they have some commons in the system structure. Fig. 2.9 shows a simplified block diagram of a typical PLC modem. It is usually composed of a power supply, a line driver, a coupling circuit, an analog frontend (AFE), a PHY controller, an application controller and diverse digital interfaces. The power supply and the signaling modules share the same interface to the PLC channel.

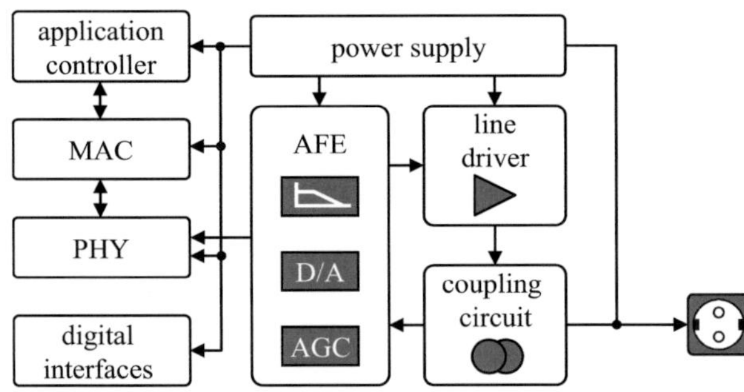

Fig. 2.9 Typical system structure of PLC modems.

The power supply module shall provide sufficient power for the internal circuitry. At the same time, it must be electromagnetically compatible. No extra interference shall be generated to degrade the quality of communication or to violate any electromagnetic compatibility (EMC) regulations. Moreover, since the power supply shares the same path with the transmitter, it must show high impedance to the transmitter, so that the transmitted signal is not attenuated by its own power supply. The PHY module manages the physical layer transmission of raw data over the specific channels. For PLC systems, the main tasks performed on the physical layer include channel coding, error detection and error correction, modulation and demodulation, bit- and/or symbol-synchronization, transmission level control and so on. Fig. 2.10 and Fig. 2.11 show PHY examples for OFDM-based PLC systems.

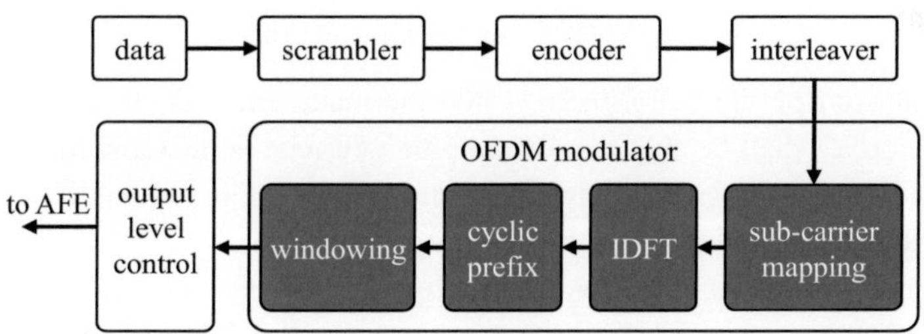

Fig. 2.10 PHY setup for an OFDM transmitter [ERD11].

In the transmitter, the payload is encoded by adding redundancy, so that the receiver can recover erroneous bits. An interleaving scheme provides diversity and reduces the correlation of received noise at the receiver. The interleaved data stream is modulated in the frequency domain and converted to OFDM symbols in the time domain. Some systems have a gain control block to scale the output signal in order to adapt to the desired transmission level. The receiver synchronizes with the transmitter, detects the symbols and carries out similar steps reversely. Sometimes blind channel estimations are made to adapt the receiver to the current channel. At the end of the receiving process the payload is obtained and is delivered to higher layers. An AFE often consists of both transmit and receive paths. It has to convert digital values to be transmitted into analog signals and to sample the received signal for conversion into digital values. Anti-aliasing and reconstruction filters are implemented, either using analog components or using digital programmable filters. The receiving path in modern AFEs is usually equipped with programmable gain amplifiers, exhibiting wide input ranges and large bandwidth to extend the dynamic range of a receiver.

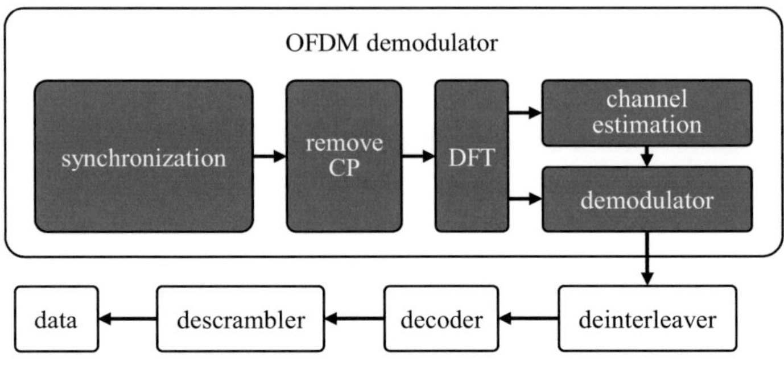

Fig. 2.11 PHY for OFDM receiver [ERD11].

The transmitted signal is "pushed" into the mains network by the line driver. The line driver shall be able to provide a high current, so that low mains impedance can also be driven properly. Most line drivers are based on AB amplifiers due to the requirement of linearity. However, high-current AB amplifiers have high power dissipation and thus temperature control and thermal shutdown are standard features to protect the devices against overheat damages.

Connecting the line driver to the mains directly will destroy the sensitive internal circuitry. The coupling circuit galvanically isolates the internal circuitry from the mains. At the same time, it blocks the mains voltage and allows the transmitted signal to pass through it. Sometimes additional coupling transformers are deployed for impedance transformation and adaptation. It also contains supplementary circuitry to protect the modem against large switching spikes, lightning strikes and overvoltage [FER10].

The media access control (MAC) module is in charge of addressing and channel access protocols for multiple-node networks. This is necessary since the powerline data link is a shared medium. All nodes in PLC networks should be addressed and coordinated, so that each node is identifiable and collisions among multiple nodes can be avoided.

The application controller is used for layers above the PHY and MAC layers. In addition, user-specific applications and management of peripherals are also handled by the controller. Any PLC modem has two fundamental interfaces: one for power supply and signaling, and the other one, also called digital interface, for digital data transfer between the modem and a PC or other devices, such as smart meters or any kind of control units. These digital interfaces, usually in the form of universal serial bus (USB), universal asynchronous receiver/transmitter (UART) or serial peripheral interface bus (SPI), are often isolated by means of opto-isolators.

2.2.1 Modulation Techniques

Modulation and demodulation techniques are the key elements of communication systems. For NB-PLC, multiple modulation schemes are involved. This section gives an overview of these different modulation techniques.

2.2.1.1 Phase-Shift Keying (PSK)

Phase-shift keying (PSK) is used in some single-carrier systems [ECH01] [ENE11]. PSK is also frequently used for sub-carrier modulation in multicarrier systems. With PSK, different zero-phase angles of the carrier are used to map the information bits. The transmitted signal $s_i(t)$ can then be expressed by

$$s_i(t) = g(t) \cdot \cos\left[\omega_c t + \frac{2\pi(i-1)}{M}\right], i = 1, 2, ...M ,\qquad(2.13)$$

where M and i refer to the total number of possible phase steps and the index of a phase value respectively. $g(t)$ is the signal pulse shape.

Since the data transmission channel can introduce arbitrary phase shifts between transmitter and receiver, the phase shift is usually dependent on the type and location of the channel. It may also vary with time and load conditions. Accurate phase tracking and compensation are needed with increasing system complexity. In addition, nonlinear operations performed by the demodulator can introduce phase ambiguity problems for the estimation of carrier phases [PRO01]. In order to overcome these difficulties, the information can be encoded in phase differences between successive symbols. For multi-carrier based modulation, such differential PSK (DPSK) can also be performed between different sub-carriers in the frequency domain [CTI11].

In a DPSK modulation scheme, the digital values are used to determine the phase shift between two symbol waveforms or two sub-carriers. Taking the DBPSK in the time domain for example, the carrier phase to convey the information bit 1 is obtained by adding 180° to the carrier phase from the previous signaling interval. If a 0-bit is sent, there is no phase change between the signals in successive signaling intervals. The receiver evaluates the phase difference between two transmitted signals over two successive signaling intervals.

Although DPSK modulation is helpful to overcome the phase ambiguity and channel-dependent phase distortions, demodulation is usually performed non-coherently, whereby tracking of the phase is not implemented. In case of non-coherent demodulation, the performance may be degraded by approximately 3 dB for M > 2. For (binary PSK) BPSK, the degradation is less than 3 dB at large signal-to-noise ratio (SNR) values [PRO01]. Of course there are also coherent demodulators for DPSK, such as those shown in [KAM11]. Nevertheless, also this kind of demodulation results in a higher probability of error than the coherent demodulation of pure PSK. The reason is that a phase error in any symbol with DPSK could lead to decoding errors spread over two consecutive symbols. This phenomenon is even obvious for BER values below 0.1.

2.2.1.2 Frequency-Shift Keying (FSK)

Conventional FSK is a non-linear modulation. The information is encoded onto carriers of different frequencies. The transmitted signal is expressed by

$$s_i(t) = A \cdot \cos\left[2\pi \cdot (f_c + m_i \cdot \frac{\Delta f}{2}) \cdot t \right], \; m_i \in \{\pm 1, \pm 3, ..., \pm(M-1)\}, \tag{2.14}$$

where A, f_c, T denote amplitude, carrier frequency, and symbol duration respectively. Δf is the frequency deviation. i is the index of a symbol.. The information bits are carried by discrete frequencies $f_c + m_i \cdot \Delta f$. The modulation index h is defined by [KAM11]

$$h = \Delta f \cdot \max\{|m_i|\} \cdot T. \tag{2.15}$$

Generally speaking, the spectral efficiency of FSK reduces with increasing h. The smallest value of h is 0.25 for minimum-shift keying (MSK) where the spectrum is used particularly efficiently. The jumping from one frequency to another in successive symbol intervals results in large out-of-band side lobes. To avoid these unwanted spectral components, it is desirable to change the carrier frequency continuously. As a result, the transmitted signal appears with a continuous phase. When FSK is modified in such a way, it is called continuous-phase FSK (CPFSK). The transmitted signal is then described by

$$s_i(t) = A \cdot \cos(2\pi \cdot f_c \cdot t + 2\pi \cdot \Delta f \cdot \int_0^t m_i(\tau)d\tau). \qquad (2.16)$$

Although FSK with low modulation index can exploit the available spectrum efficiently, it is sensitive to narrowband interferers. To improve robustness against this kind of noise, spread FSK (S-FSK) has been developed which introduces the advantages of spread spectrum technology into classical FSK systems. S-FSK technique is based on binary FSK and standardized in IEC61334-5-1 for PLC applications. According to this standard, the carriers for 0-bit and 1-bit are called space and mark respectively. The frequency deviation is recommended to be larger than 10 kHz. The receiver performs conventional FSK demodulation at the two expected frequencies. Furthermore, the quality of both demodulated signals, denoted by d_S and d_M, is measured and observed. If both signals show similar strength, then the decision unit will deliver bit 0. Otherwise, if d_S is larger than d_M, then bit 1 is delivered. However, if the SNR at one carrier is much better than that at the other one, the decision unit will solely regard the better signal and compare it with a threshold for bit decision.

2.2.1.3 Code Shift Keying (CSK)

Code shift keying (CSK) is a kind of spread spectrum modulation technique [JOE09]. The information bits are mapped onto cyclic time shifts of a basic waveform. In principle, the signal can take any waveform whose autocorrelation function exhibits a large and sharp main lobe. Assuming that the data stream is transmitted in blocks of N bits, then $M=2^N$ symbols are used. There are M possible positions for the cyclic rotation. The modulation can be implemented either non-differentially or differentially. In a non-differential CSK, the positions of the basic waveform is used as a reference and the rotations of symbols are used to encode the data. The differential CSK (DCSK) utilizes the relative rotation between symbols in successive signaling intervals to convey data. At the receiver, the basic waveform is saved as a template. The received signal is fed into a ring buffer and circularly rotated. After each rotation, the cross-correlation of the template with the rotated waveform is calculated by a correlator. At the end, the expected rotation corresponds to the maximum correlation value. The non-differential CSK uses this detected rotation amount to decode the data. For DCSK two successive rotation steps are needed to determine the difference which is used to decode the data [YIT01].

2.2.1.4 Orthogonal Frequency Division Multiplexing (OFDM)

OFDM is widely used in both wireless and wireline communications, not only because of its high spectral efficiency, but also due to its robustness against multi-path fading and narrowband interferers [EDF96]. The basic idea of OFDM is to divide the transmission band into N_c sub-channels. The data to be transmitted is distributed to these sub-carriers, each of which can be modulated separately. An OFDM symbol is adding the waveforms of all sub-carriers together. Each waveform has duration T_s and T_s is therefore the OFDM symbol duration. The typical OFDM symbol duration is relatively long in comparison with conventional single-carrier modulation techniques. Therefore, OFDM systems are more robust against channel fading, impulsive interference and imperfect symbol synchronization. Although a longer symbol results in lower data rate on each sub-carrier, the overall data rate is actually increased due to a high number of sub-carriers. To avoid inter-carrier interference, the sub-carriers exhibit orthogonality by a proper assignment of frequency spacing and symbol duration.

Typical single-carrier phase modulation schemes can be applied to individual sub-carriers. The DPSK is supposed to be one of the most popular modulation techniques due to the robustness against a sampling frequency offset. It can be made either in time or frequency domains. In the frequency domain DPSK, the phase difference between two sub-carriers is used to carry information. The first sub-carrier is usually used as a phase reference for other sub-carriers. In the time domain DPSK, the phase reference is provided by the first OFDM symbol. For each sub-carrier, the phase difference between two symbols is used to carry information.

The fundamental structures of an OFDM modulator and a demodulator have been shown in Fig. 2.10 and Fig. 2.11 respectively. At the transmitter, the serial source symbols at the input of the "sub-carrier mapping" block are obtained after source scrambling, channel coding and interleaving. Let us assume that the bit rate of the serial source is R_d, and total N_d data symbols shall be conveyed within one OFDM symbol. Then the OFDM symbol duration must be

$$T_s = \frac{N_d}{R_d}.\tag{2.17}$$

In order to fulfill the orthogonality requirement for the sub-carriers, the carrier frequency spacing is

$$f_\Delta = \frac{1}{T_s}.\tag{2.18}$$

The mapped and serial-to-parallel converted source data stream, denoted by S_i, $i=0,\ldots,N_c\text{-}1$ is transformed from the frequency domain into the time domain by an IDFT:

$$s(t) = \frac{1}{N_c} \cdot \sum_{i=0}^{N_c-1} S_i \cdot e^{j2\pi \cdot i \cdot f_\Delta \cdot t}.\tag{2.19}$$

After the IDFT, a guard interval is attached to the OFDM symbol signal in the time domain. In wireless or BB-PLC, the guard interval is used to eliminate inter-symbol interference (ISI) caused by multipath distortion. In a NB-PLC channel, ISI can also be caused by synchronization jitter between transmitter and receiver. In this case, the length of the guard interval should be larger than the maximum jitter error [KIS08]. At the receiver, the guard interval is usually removed from the received signal. The rest of the received signal will be converted into the frequency domain using a DFT. In the case of symbol synchronization using the mains zero-crossing detection method, the guard interval must be four times larger than the maximum synchronization deviation, and the demodulation at the receiver shall be delayed by half guard interval length to eliminate ISI [KIS08]. Demodulation must be made for each sub-carrier separately. Additional channel estimation and frequency synchronization can be provided to improve the transmission quality. An important prerequisite for an OFDM modulation scheme to work is that the time variance of the channel is negligibly small within an OFDM symbol [FAZ03]. Another challenging topic in system design is how to reduce the peak-to-average power ratio (PAPR) which is given by

$$PAPR = \frac{\max\{|s(n)|^2\}}{\frac{1}{N_c} \cdot \sum_{n=0}^{N_c-1} |s(n)|^2}.$$

(2.20)

PAPR of a single frequency waveform, for example a sinusoid is 2. For an OFDM signal, suppose that all sub-carriers have the same information bits, then we have

$$S_i = S_j, \quad i, j \in \{0, N_c - 1\},$$

(2.21)

Considering the waveform given in (2.19), the maximum power is

$$P_{\max} = \max\left[\frac{1}{N_c} \cdot \sum_{i=0}^{N_c-1} S_i \cdot e^{j2\pi \cdot i \cdot f_\Delta \cdot t} \cdot \frac{1}{N_c} \cdot \sum_{i=0}^{N_c-1} S_i^* \cdot e^{-j2\pi \cdot i \cdot f_\Delta \cdot t} \right] = \frac{|S|^2}{N_c}.$$

(2.22)

The average power is

$$P_{average} = \frac{1}{N_c} \cdot \frac{|S|^2}{N_c} = \frac{|S|^2}{N_c^2}.$$

(2.23)

As a result, the PAPR can reach N_c. This high PAPR challenges the line driver at the transmitter and the operational amplifier at the receiver in that way that both of them must provide large dynamic ranges. Otherwise nonlinear distortion will occur. As a result of the non-linear distortion, performance degrades and the out-of-band power will be enhanced. There are various methods to reduce the PAPR of OFDM systems, for example clipping, coding, peak windowing and so on. However, most of them can only provide limited success with high complexity, large overhead and usually a degradation of communication performance [FAZ03].

2.2.2 Standardization and Available PLC Systems

Till now, two generations of NB-PLC have been developed. The first generation is based on simple single-carrier or dual-carrier techniques. Representative standards include LonWorks® (ISO/IEC14908-3), KNX (ISO/IEC14543-3-5), CEBus (CEA-600.31), IEC61334-3-1, and IEC 61334-5. Some technologies such as Meters & More, Ariane Controls, BacNet and HomePlug C&C are also widely deployed, although they are not ratified by international standard development organizations (SDO). The second generation utilizes OFDM to achieve higher data rates. Early systems were reported in [DEI99] [GAL01]. Powerline intelligent metering evolution (PRIME) and G3 are developed within the scope of the European project "OPEN Meter" [OKS11]. In the following sections, some representative technologies and systems will be discussed in detail.

2.2.2.1 LonWorks and BPSK

LonWorks stands for local operation network. Developed since 1988, it is a platform for implementing control network systems. This platform contains several components: smart transceivers, development tools, routers, network interfaces and protocols, and internet servers [ECH01]. Four parts of the LonWorks® technology have been granted by international organization for standardization (ISO) and international electrotechnical commission (IEC) as open standards.

Table 2.1 LonWorks® Standards

Parts	Contents
ISO/IEC14908-1	Control network protocol
ISO/IEC14908-2	Twisted-pair wire signaling technology
ISO/IEC14908-3	Powerline signaling technology
ISO/IEC14908-4	IP compatibility technology

LonWorks has defined several common "channel types" in which the technology can be used. Among these channels, three types are defined for powerlines: PL-20A, PL-20C and PL-20N, corresponding to the CENELEC A-band, C-band with access control, and C-band without access control respectively. It is worth noting that these "channel types" do not describe the channel characteristics. Instead, they are only PHY requirements for communication systems shown in Ta-

Table 2.2 PL-20 Powerline Specifications [ECH01]

Parameters	PL-20A(L-N)	PL-20(L-N)	PL-20(L-E)
Coupling	Line-to-neutral	Line-to-neutral	Line-to-earth
Bit rate (kbps)	3.6	5	5
Modulation	BPSK		
Frequency band (kHz)	70-95	125-140	
Output level	$\leqslant 116\mathrm{dB\mu V}$ for class 116 EN50065-1 $\geqslant 115\mathrm{dB\mu V}$ otherwise		

ble 2.2. For the PLC channels there are standard transceivers such as PLT-20, 21, and 22 as well as smart transceivers such as PL3120, PL3150 and PL3170.

A smart transceiver is a system-on-chip (SoC) which is composed of a stand-alone microprocessor core, also called neuron core, and communication transceivers. PLT-20 and PLT-21 use conventional single-carrier BPSK modulation which is sensitive to narrowband jammers. To improve the immunity, a second carrier is added in the standard transceiver PLT-22 and all of the smart transceivers. Carriers are centered at 75 kHz and 86 kHz for CENELEC A-band, or 115 kHz for B-band and 132 kHz for C-band. Each data transmission begins on the primary carrier. Since PLT-20 and 21 also use the primary frequency, they can communicate with devices that are based on the PLT-22 transceiver. If the primary carrier is blocked by noise, the transceivers are able to switch to the second carrier. Since the probability that two carriers are blocked at the same time by narrowband interferers is relatively low, the immunity can be increased [ECH02]. In addition to the BPSK modulation, the manufacturer has also released the transceiver PLC-30 which is based on direct sequence spread spectrum technique (DSSS). It provides a data rate of 2 kbps in the frequency band 9-95 kHz. However, despite the lower data rate, the system based on the PLT-30 transceiver shows much lower performance than the one based on BPSK during the tests made by the manufacturer [SUT98].

2.2.2.2 Meters and More

Meters & More technology is a complete solution for automatic meter reading applications. It is a single carrier system and is proposed as a standard within the scope of Open Meter project. Its PHY layer is using DBPSK for payload and non-differential BPSK for the frame header. The data rate can reach 9600 bps without channel coding and 4800 bps with a convolutional code. The coding is applied on the payload only, whilst the header is left uncoded. The carrier frequency is centered at 86 kHz. Additional features are implemented on PHY layer to improve the system performance. In order to synchronize the clocks and optimize the bit sampling timing, the transmitter sends a sequence of alternating 1- and 0-bits as a preamble of one frame. The phase-locked loop (PLL) at the receiver can adjust itself to lock the best position to sample symbols. To prevent cross-talk among different line phases, the frame transmission occurs by devices - including repeaters and target nodes - on the same mains phase [ENE11]. A representative device which is compatible with the Meters & More technology is the ST7580 system on chip (SoC) [STM09].

2.2.2.3 FSK and Spread-FSK

Binary FSK systems are also widely used, for example ST7538, ST7540 [ST7538] [ST7540]. The ST7538 provides half-duplex communication. The carrier frequency is programmable to 8 values: 60, 66, 72, 76, 82.05 and 86 kHz for the CENELEC-A band, 110 kHz for the B band and 132.5 kHz for the C band. Four baud rates - 600, 1200, 2400 and 4800 bps - are achievable. Based on the ST7538 core, the ST7540 transceiver is designed to achieve reduced size and cost. Besides PSK modulation for the Meters and More technology, the ST7580 also supports binary-FSK modulation. [CYP11] has also introduced a programmable CY8CPLCx0 family for NB-PLC in March 2010. This PLC family is based on BFSK modulation, providing half-duplex communication with variable data rates below 2400 bps. The first S-FSK modem AMIS-30585 can provide 1.2 kbps for 50 Hz mains networks. Its second solution AMIS-49587 is equipped with an ARM7 microprocessor core, including both the PHY and MAC layers. The data rate has been raised to 2.4 kbps (50 Hz). In 2012, the third generation NCN49597 SoC was brought onto the market. It is based on a 32-bit ARM Cortex M0 processor core and can reach a data rate of 4.8 kbps (50 Hz).

2.2.2.4 Differential Code Shift Keying (DCSK)

The first DCSK modem was IT800 PLC SoC, which contains modules for both PHY and DLL. There is no integrated application controller on the chip. Thus, an external host must be used to manage the data flow. The maximum raw data rate can reach 7.5, 5 and 1.25 kbps in standard, robust and extremely robust mode, respectively. For the CENELEC band the data rates in the robust and extremely robust modes are 2.5 and 0.625 kbps, respectively. The PHY layer has a 10-bit DAC for the transmitting path. The reception path contains three 1-bit ADCs. Three external bandpass filters are needed to provide the ADCs with the corresponding input signals. The demodulation is based on 1-bit correlation. IT700 is a highly integrated PLC SoC containing PHY, DLL and a host 8051 controller for the protocol stack and applications. The maximum data rate reaches 7.5 kbps. It is designed for HomePlug Command & Control and complies with worldwide regulations, for example, Federal Communications Commission (FCC) part 15, Association of Radio Industries and Businesses (ARIB) and CENELEC bands. In addition, a 13-bit ADC replaces the 1-bit ADCs for the reception path. IT900 DCSK Turbo IC is the next generation of IT700 with speeds of up to 500 kbps and support for SE 1.0, SE 2.0, IPv4 and IPv6.

2.2.2.5 OFDM

In the second generation of NB-PLC systems, PRIME and G3 are two specifications proposed within the scope of the European Open Meter project. Both specifications are based on OFDM. The PRIME technology uses 97 sub-carriers with a spacing of 488.28 Hz. The spectrum of the transmitted signal extends from 42 to 89 kHz. Data can be modulated using DBPSK, DQPSK or D8PSK in the frequency domain. A chirp waveform is placed at the start of a frame for synchronization purposes. 13 pilot sub-carriers are used in the header symbols to estimate the sampling start error and the frequency offset. In subsequent payload symbols, the first pilot sub-carrier is used as a phase reference and the others are used as normal sub-carriers for data bits. Convolutional coding with a code rate of 0.5 is used for forward error correction. The interleaving is made per OFDM symbol. Within each symbol, the adjacent coded bits are distributed to non-adjacent sub-carriers so that the occurrence of errors can be scattered randomly instead of appearing in a burst manner. Due to the large number of sub-carriers the PRIME technology can deliver maximum raw data rates up to 128.6 kbps [CTI11].

The G3 technology deploys 36 sub-carriers. The frequency spacing is about 1.56 kHz. G3 has raised its sampling rate to 400 kHz, more than 4 times above the upper bound of the CENELEC-A band. The reason is to provide some margin for signal filtering. Data is modulated onto individual sub-carriers using either DBPSK or DQPSK in the time domain. The symbol synchronization is based on a sequence of specific OFDM symbols. Due to the reduced number of sub-carriers and modulation index, G3 can deliver a raw data rate up to 33.4 kbps. The data stream is also protected by a convolutional code with the rate 0.5. In addition, data is repeated several times for the frame control header (FCH), and when the device is in robust mode. Non-FCH data are encoded with a shortened Reed Solomon code too. The interleaving is made prior to DBPSK modulation [HOC11]. Additional features are also implemented in G3, for example, a pre-emphasis is made for the transmitted signal, so that the received signal has a flat spectrum and the average power on each active sub-carrier is kept within ±2 dB of the overall average power. Channel adaptations, such as the selection of available sub-carriers, as well as the decision of optimum modulation schemes are made according to the measured quality of received signals. A G3 system has the capability of identifying the mains phases. Moreover, cohabitation with S-FSK systems can be achieved by inserting deep notches. A summary of the parameter comparison between PRIME and G3 technology is shown in Table 2.3.

ST7590 is the first complete OFDM SoC for NB-PLC. It has been introduced and applied in advanced automatic meter management equipment in Spain since September 2010 [ST7590]. This SoC includes a PRIME compliant PHY layer, a fully integrated networking module, as well as an AFE and line driver. Later more and more SoC and application-specific integrated circuit (ASIC) have been released for AMR applications, for example MB87S2080 [FUJ10], ADD1021 [ADD1021] and ADD1022 [ADD1022]. With respect to the G3 technology, [MAX08] has introduced an AFE chip MAX2990 in 2008. Based on the MAX2990, a PLC modem MAX2992 has been developed in 2011, and it became the first G3-PLC chipset solution.

As a result of the Open Meter project, in total three specifications or standards are proposed. Instead of focusing to one technology, the semiconductor industry is delivering a variety of flexible and programmable solutions. For example, a flexible PLC development platform named "plcSUITE" has been released which is composed of a C2000 microcontroller unit (MCU), a separate AFE and docking station modules. The platform supports IEC 61334 (S-FSK), G3 and PRIME technologies [TEX11]. Although this platform is not suitable for applications in real smart meters, it is supposed to be a convenient development environment

Table 2.3 Parameter Comparison between PRIME and G3 for CENELEC

Parameters	PRIME	G3
Frequency range (kHz)	41.992 – 88.867	35.938 – 90.625
Baseband clock (kHz)	250	400
Subcarrier spacing (Hz)	488.281	1562.5
Symbol interval (µs)	2240	-
Cyclic prefix (µs)	192	75
Number of carriers	97	36
M-ary QPSK	M = 2,4,8, differential encoding in frequency	M = 2,4, 8 differential encoding in time
Maximum raw data rate (kbps)	128.6	48
Error correction	Convolutional code	Reed Solomon code, Convolutional code, Repetition code
Interleaving	Per OFDM symbol	Per data packet

which allows the developers to optimize their OFDM implementation for specific transmission channels. Once a design is finished, developers can integrate their optimized design into a single hardware solution [MON10]. Similarly, a SoC named STarGRID is a flexible and scalable SoC which supports PRIME, Meters & More and IEC61334-5-1 [STarGRID].

3 Narrowband-PLC Testbed

It has been shown in the previous chapter that the NB-PLC covers a wide range of technologies and system structures. Therefore, a universal testbed is desirable for the evaluation of these different systems. As mentioned above, this testbed is based on a channel emulator which mimics the communication-related channel behavior. Additional components must be used to provide a real-world mains environment. In order to give a quick impression of the overall construction and a better understanding of the role that the channel emulator plays, this chapter avoids detailed insights into the channel emulation algorithms. Instead, it begins with a brief overview of the testbed. Key components and the signal flow within the testbed will be introduced. After that, the modular hardware design of a unidirectional emulator will be illustrated, followed by the realization of bidirectional channels using two unidirectional emulators. Special components that are needed for the construction of a LV mains environment will also be analyzed.

3.1 Overview of the Testbed

Fig. 3.1 shows the block diagram of the testbed. It consists of a test server PC, a unidirectional channel emulator, two coupling circuits (CC), two line impedance stabilization networks (LISN), and an uninterruptible power supply (UPS). DUTs are connected to the Tx/Rx electric sockets P_2 and P_7 respectively. At the same time, the DUTs are connected to the test server via paths P_1 and P_8. The test server configures the DUTs, controls the emulator and manages the whole test process. The UPS provides an equivalent mains voltage to supply the emulator and the DUTs. The UPS also isolates the test environment from the mains network. Each of the LISNs provides well-defined stable impedance to the test environment and prevents the transmitted signal from reaching a receiver via the mains path P_9-P_{10}-P_{11}. Unwanted high frequency noise generated by the UPS can also be blocked by the LISNs. The coupling circuits CC and CC' are necessary to separate the emulator from the mains voltage and to exchange the "pure" communication signals between the emulator and the DUTs. From the signal/data flow's point of view, a closed loop is formed within the testbed: information bits, also called test patterns, are generated by the test server and transferred to Tx via P_1. Tx modulates and converts the digital values to a signal waveform and sends this to the emulator via P_3 and P_4. The emulator attenuates the signal and adds artificial noise to it, just like the real PLC channel would do. The distorted signal reaches Rx via P_5, P_6 and P_7, and the information bits are demodulated and sent back to the test server over P_8. This closed loop enables convenient estimation of the BER and makes automated test procedures possible.

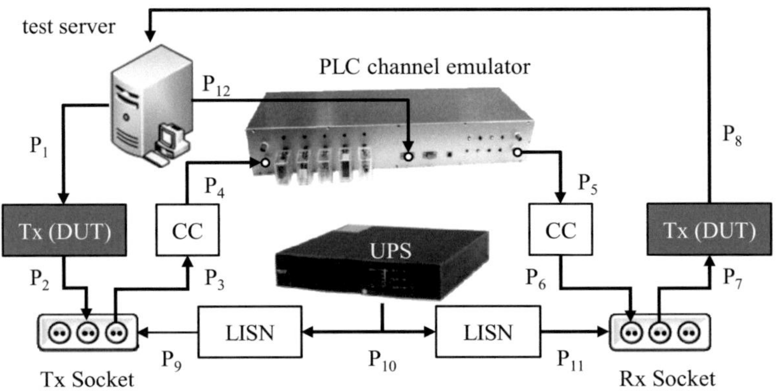

Fig. 3.1 Block diagram of the testbed.

3.2 NB-PLC Channel Emulator

The channel emulator is the key component of the testbed. Its block diagram is shown in Fig. 3.2. The design is composed of four hardware modules. The input module emulates the access impedance with digitally switchable passive RCL-networks. The AFE contains a 12-bit ADC and two 14-bit digital-to-analog converters (DAC). The sampling rates of the ADC and the DACs are configurable. Two additional digitally controlled attenuators are utilized to achieve a wide range for SNR adjustment. Each attenuator has a dynamic range of more than 75 dB with a resolution of 0.37 dB. The FPGA is the core component for realizing signal processing and control algorithms. It implements digital filters and stores the impulse response values for up to 10 different channels in a filter coefficient memory. The filter can be reloaded with a new set of coefficients within 10.5 microseconds. The noise unit reproduces complete noise scenarios. The time variance control unit manages the change of states for the access impedance, digital filters and noise. The event table holds the time schedule for each switching event. The SNR unit determines the attenuation values for the signal and the noise paths. The output module contains a high power amplifier which has very low output impedance and can deliver a current up to 1.5 A. This module also scales the output voltage to an appropriate level. The emulator is connected to the test server via a standard serial interface so that the emulation process can be controlled externally.

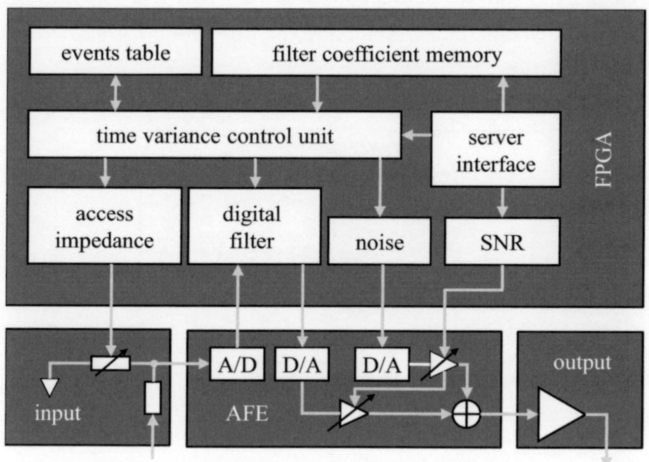

Fig. 3.2 Block diagram of the emulator.

3.3 Bidirectional PLC Channel Emulation

Real-world PLC applications usually include two-way data transmission. Many PLC systems use error control techniques, such as automatic repeat request (ARQ) to guarantee error-free data transmissions. As ARQ is based on feedback from the receiver, bidirectional communication is a prerequisite. Furthermore, some systems are able to estimate the channel conditions, and adjust important parameters such as carrier frequencies or the modulation scheme to achieve better communication quality. In general, such systems can detect changes of channel characteristics, and thus are able to adapt themselves to the new channel conditions. All these advanced features are based on channel measurements for both directions. Therefore, it is expected that the testbed can emulate bidirectional data transfer links as well.

3.3.1 Extended Testbed for Bidirectional Channel Emulation

It is possible to deploy two unidirectional emulators to realize a bidirectional testbed. As shown in Fig. 3.3, a second emulator is used for the return link. In addition, several modifications have to be done to ensure the expected function. Recall that the line driver has very low output impedance. It makes the equivalent impedance of the CC' on the mains side also very low. As a result, the transmitting DUT faces a low-impedance load, and the transmitted signal will suffer from an unwanted attenuation.

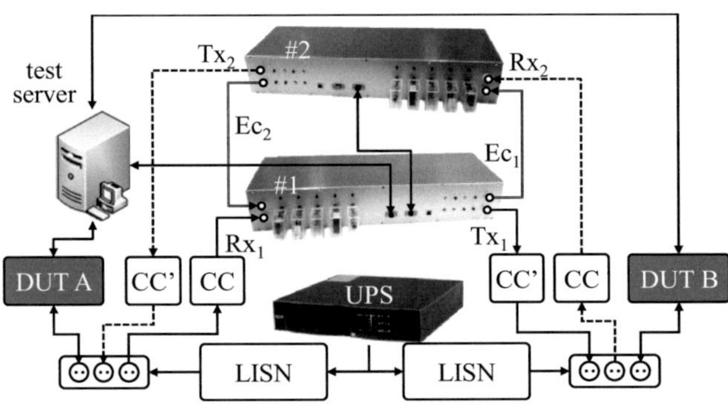

Fig. 3.3 Block diagram of a bidirectional emulator.

To overcome this problem, a resistor between 10 to 20 Ω can be connected in series to the blocking capacitor in both CC's. This resistor shall be able to sustain 5 watts, so that it will not be destroyed e.g. by emulated high-level noise. Although the series resistor introduces additional attenuation of the emulated signal and noise, the attenuation can be compensated by the line driver with a proper pre-amplification factor. Control codes from the test server can be sent to both emulators via a cascaded command bus. Furthermore, two paths, denoted by Ec$_1$ and Ec$_2$ are introduced to cancel unwanted echoes which are caused by the looped connection. Details of the echo cancellation will be introduced in the following section.

3.3.2 Unwanted Echo Effect

Although the testbed can be extended with relatively low complexity from the hardware point of view, the closed-loop connection, formed by emulator #1 $\rightarrow$ $Tx_1 \rightarrow CC' \rightarrow CC \rightarrow Rx_2 \rightarrow$ emulator #2 $\rightarrow Tx_2 \rightarrow CC' \rightarrow CC \rightarrow Rx_1 \rightarrow$ emulator #1, introduces an unwanted echo effect. Let us assume that DUT-A transmits continuously. Then the delayed and attenuated transmitted signal reaches DUT-B via the first emulator, and at the same time it goes back to DUT-A over the second emulator and overlaps with the original transmitted signal. In this way, artificial echoes occur and degrade the test accuracy. The same happens if DUT-B transmits and DUT-A receives. Fig. 3.4 illustrates an example of the echo in the time domain. y_B and y_A are the waveforms measured on the sides of DUT-B and DUT-A respectively. An impulsive waveform has been generated by emulator #1 at t_1. It is picked up and distorted by emulator #2 and appears at DUT-A at t_2. Again, emulator #1 gets the modified waveform and sends it to DUT-B at t_3. This process repeats again and again, until the impulses "disappear" after several turns. It can also be observed that each emulator introduces an intrinsic delay of about 250 µs into the signal propagation. This delay can be adjusted arbitrarily and will be discussed later in chapter 4.

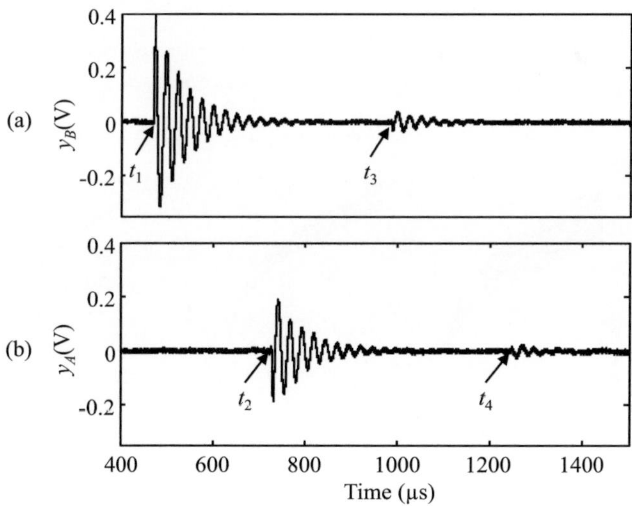

Fig. 3.4 Example of an echo in the time domain.

Besides the time domain, the influence of the echo effect on the magnitude of the frequency response is also investigated. First, both emulators get the same channel transfer function (CTF), and the intrinsic delay is set to 100μs for both emulators. Let H_{21} and H_{12} denote the CTFs for the emulator #1 and #2 respectively. H_{21} doesn't change in the following observation, while additional flat attenuation is added to H_{12} step by step. The actual emulated CTFs are measured using a vector network analyzer in each step. Fig. 3.5 shows three pairs of expected and actually measured CTFs. In total 6 subplots are depicted in two rows, and each row contains 3 subplots. The first row shows the expected and the actual magnitude responses for $|H_{21}|$ and the second row is for $|H_{12}|$. The expected magnitude responses are denoted by (e) in all plots, and the actual responses are denoted by (a), (b) and (c) respectively. Let $|H_{12}|$-(a) denote the curve (a) of $|H_{12}|$. It can be seen that both $|H_{21}|$-(a) and $|H_{12}|$-(a) differ from the expected (e) curves.

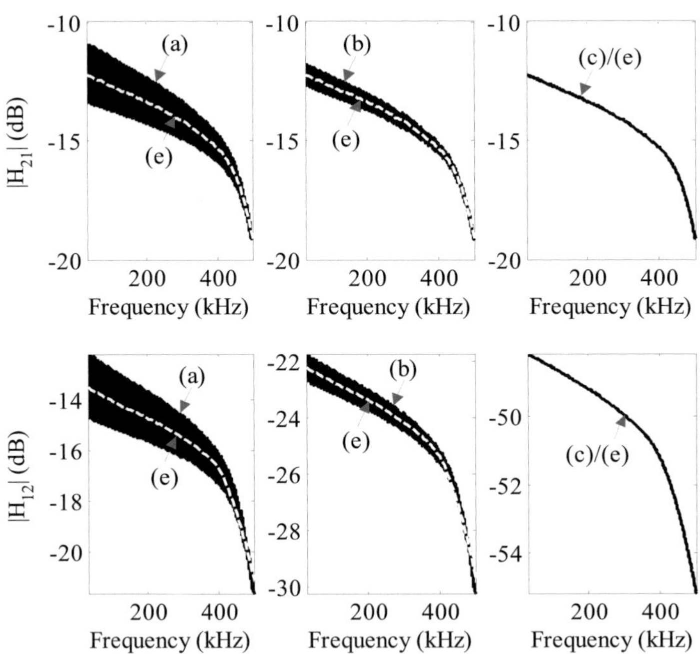

Fig. 3.5 Expected and measured magnitude responses. Curves (a), (b) and (c) are CTFs suffering from the echo effect. (e) is the expected CTF.

Fig. 3.6 shows a zoomed view of |H21|-(a) in the frequency range 30-45 kHz. Obviously there is an oscillation following a sinusoidal function of frequency, and the oscillation reaches local maxima at distances of 5 kHz. Additional investigations have shown that this distance is inversely proportional to the echo round trip delay. The superimposed waveform measured at DUT-B is

$$y_{Bn} = \sum_{m=1}^{n} \sin\left[\omega \cdot t + \omega \cdot (m \cdot \tau_1 + m \cdot \tau_2 - \tau_2)\right], \tag{3.1}$$

where n is the number of returns. ω is the angular frequency, τ_1 and τ_2 are delays from emulator #1 to #2 and from #2 to #1 respectively. Limiting n to 2, the waveform can be approximated by

$$y_{B2} = 2 \cdot \sin\left[\omega \cdot \left(t + \frac{3}{2}\tau_1 + \frac{1}{2}\tau_2\right)\right] \cdot \cos\left(\omega \cdot \frac{\tau_1 + \tau_2}{2}\right). \tag{3.2}$$

The cosine-term determines the sinusoidal fluctuation over frequency. The periodicity is the reciprocal value of half round trip delay.

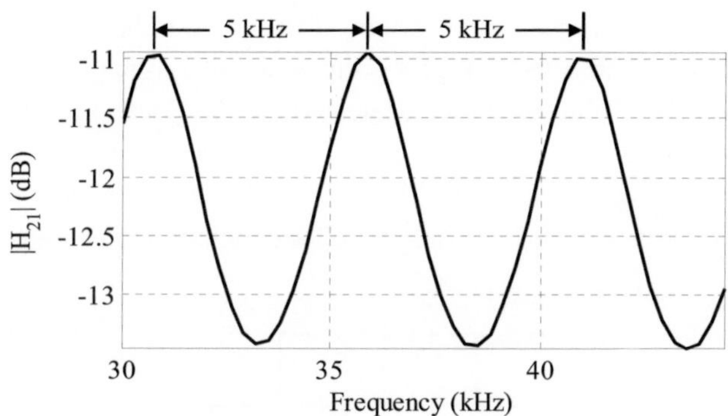

Fig. 3.6 Zoomed magnitude response for |H21|.

In the next step, the attenuation of $|H_{12}|$ is increased by 8.56 dB. The local maxima of the sinusoidal traces $|H_{21}|$-(b) and $|H_{12}|$-(b) are reduced accordingly. As the attenuation further increases, the magnitude deviation becomes smaller. The oscillations are no longer visible in $|H_{21}|$-(c) and $|H_{12}|$-(c).

The echo-problem can be solved by either driving the testbed in a half-duplex mode or using echo-cancellation methods. The half-duplex approach takes advantage of the fact that most NB-PLC systems are designed for a master-slave communication. The master initiates the data transfer, and slaves are not allowed to transmit without a request sent by the master. In this case, signal detection algorithms can be implemented in both emulators. As soon as a data transmission in one direction has been detected, the signal path in the other direction will be broken. In situations in which full-duplex channels are desired, for example, in the evaluation of medium access control algorithms, hybrid circuits or directional couplers could be used to separate the signal in one direction from the one in the opposite direction. A prerequisite for this method to work is accurate impedance matching. However, it is difficult to match the equivalent impedance of PLC systems over the whole frequency range. Furthermore, there are no exact specifications for coupling circuits and AFEs of NB-PLC systems. Different NB-PLC modems can have different output impedance. It is time consuming to match the impedance for each DUT. Therefore, the application of hybrid circuits or directional couplers alone cannot solve the problem.

3.3.3 Echo Cancellation using Digital Filters

Similar echo problems appear in telephony. Many efforts have been made to eliminate undesired echo effects. Fig. 3.7 illustrates the principle of a conventional echo canceller. The signal x_1 generated by the local transmitter goes through an unexpected echo path. As a result, an undesired echo signal e_1 is superimposed with the distant transmitted signal x_2. The primary job of the echo canceller is to remove e_1 from x'_2. The canceller simulates the characteristics of

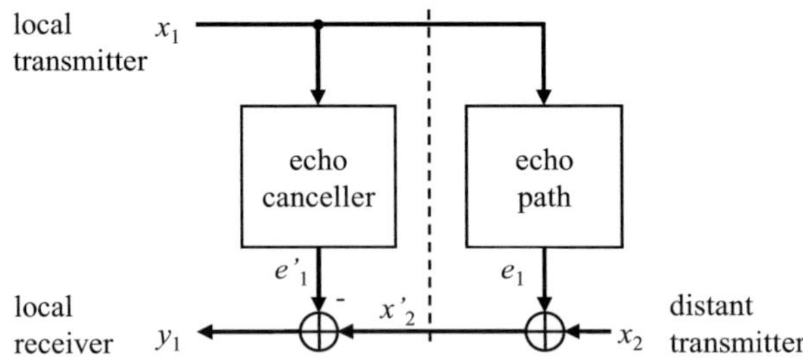

Fig. 3.7 Principle of an echo canceller.

the path from the local transmitter to the distant transmitter and reproduces a copy of the echo e'_1. This copy is subtracted from x'_2 to yield y_1, which ideally contains the distant transmitted signal x_2 alone. The echo canceller is usually implemented in discrete form as a digital filter [BAR04].

The possibility of implementing the echo cancellation with a simple digital filter encourages its application in the testbed. Fig. 3.8 shows a simplified block diagram which illustrates the specific situation in the testbed. Suppose x_1 is a sum of the emulated noise and the distorted transmit signal. It is scaled by the line driver inside emulator #1 (x_2) and is delivered to DUT B via the coupling circuit CC (x_3). The signal level of x_3 is determined by CC and the equivalent impedance Z_B of DUT B. Let x_B be the transmitted signal generated by DUT B. It is superimposed with x_3 and the sum x_4 is picked up by CC'. Till now, the echo of x_1 enters

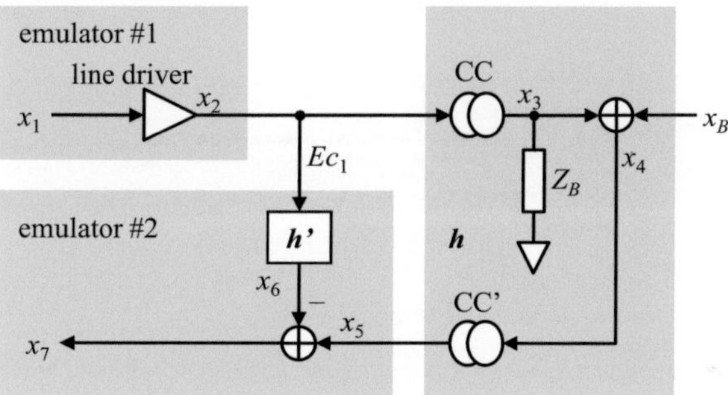

Fig. 3.8 Block diagram of echo cancellation for the testbed.

the emulator #2 via its echo path CC → DUT B → CC'. Let h denote the impulse response of this echo path. It is known to emulator #2 and is simulated by h'. Meanwhile, x_2 is fed to emulator #2 via the aforementioned echo-cancellation path Ec_1. The copy of the undesired echo, denoted by x_6, is generated by convoluting x_2 with h'. Finally x_6 is subtracted from x_5 and x_7 is supposed to be echo-free and contains x_B only. In principle, h depends on Z_B and will change if a different DUT is connected. In this case, it is necessary for emulator #2 to estimate the new h. The estimation procedure is quite simple. Emulator #2 triggers emulator #1 to generate a single impulse x_1. At the same time, emulator #2 starts to record x_5 at its input. The acquired waveform is the desired impulse response h'.

Fig. 3.9 shows magnitude responses before and after the echo cancellation. Originally the channel has the same conditions as the curves (a) in Fig. 3.5. After having applied the echo cancellation approach, the echo has been reduced to a great extent in (a'). It can also be seen that there is still a residual oscillation in (a'). This error could be caused by the estimation error of h as well as the rounding error occurring in the digital filter. However, the absolute level of the oscillation falls below 0.2 dB. Therefore, the residual echo effect will no longer cause significant degradation.

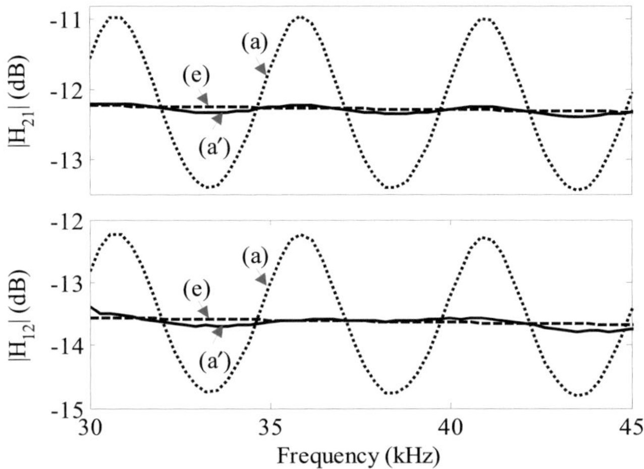

Fig. 3.9 Results of echo cancellation: (a) magnitude responses before echo cancellation (a') magnitude responses after echo cancellation (e) expected magnitude responses.

3.4 Line Impedance Stabilization Networks (LISN)

Selection and design of LISNs are also critical and time-consuming tasks in the construction of the testbed. Improper LISNs cannot guarantee satisfying performance. Even worse, it could generate noise by itself which disturbs the test process and render all effort in the channel emulator useless. There are off-the-shelf LISN circuits that can be obtained commercially. Two specific LISNs have also been implemented in the scope of this work. A comparison between the off-the-shelf and the specific LISNs is made. In the first specific LISN circuit, toroid cores are used to construct inductors. Despite all advantages which toroid cores can provide, unwanted impulsive noise is observed. The noise level depends on the equivalent impedance of the connected DUTs. The noise origin is analyzed both theoretically and experimentally. The unwanted noise is eliminated in the second solution.

3.4.1 Off-the-Shelf LISN

The commercially available LISN is often used in electromagnetic interference testing. It allows the measurement of conducted radio frequency (RF) interference from active electrical appliances. Each LISN must provide the measurement receiver (50 Ω input impedance) with stable and defined impedance. It should provide coupled RF interference and block the 50/60 Hz mains voltage at the interface to which a standard electromagnetic interference (EMI) receiver (e.g. spectrum analyzer) is connected.

A LISN can be constructed in different ways. EN50065-1 deploys a LISN for interference measurements [CEN01]. Its schematic diagram is shown in Fig. 3.10. The circuitry is composed of two mirrored sections, each of which is a fourth-order LC lowpass filter with a series branch. The inductors L_1 and L_2 block HF interference from the mains while the capacitors C_1 and C_2 provide alternate paths for noise currents. The capacitor C_3 blocks DC and 50 Hz voltages for the EMI receiver. The resistor R_3 discharges C_3 in the case that no receiver is connected. R_I denotes the input resistance of the EMI receiver or a dummy load. The dummy load has the same value as the input impedance of the receiver. If the receiver is connected to one section, a dummy load must be connected to the other section, so that the impedances between L and PE and between N and PE are the same. Table 3.1 shows two configurations for different operating frequencies. In order to examine the possibility of applying the LISN in our test platform, simu-

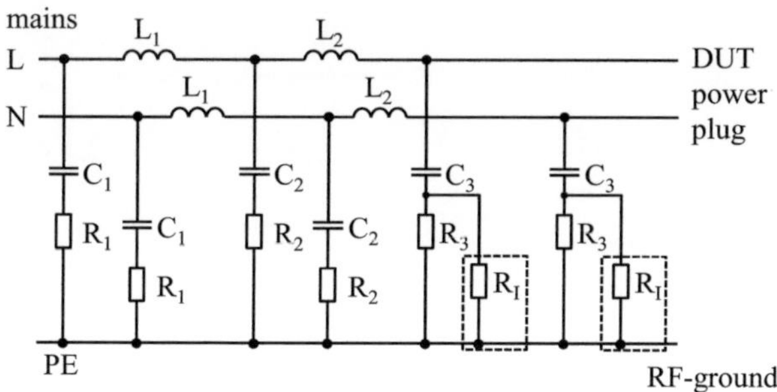

Fig. 3.10 Schematic diagram of a LISN defined in EN50065-1.

lations are made to for the transfer characteristics for two directions: from mains to DUT and from DUT back to mains. Despite differences in the component values, both configurations have almost the same transfer function for each direction in the frequency range of interest. In Fig. 3.12, the dotted line denoted by "mains-DUT, EN-1" is the transfer function for the direction from mains to DUT, while the "DUT-mains, EN-1" is for the opposite direction. As mentioned before, the attenuation caused by the two LISNs shall be higher than the maximal value of 75 dB. Unfortunately, the standard LISNs can provide only 40 dB at 30 kHz and 70 dB at 100 kHz for the test platform. This will degrade the test quality in the CENELEC band.

Table 3.1 Components of the LISN Defined in EN50065-1

Configuration	Frequency	L_1 (μH)	L_2 (μH)	C_1 (μF)	R_1 (Ω)	C_2 (μF)	R_2 (Ω)	C_3 (μF)	R_3 (kΩ)	R_I (Ω)
EN-1	9-95 kHz	250	50	4	10	8	5	3.3	1	50
EN-2	95 kHz-30MHz	250	50	4	10	8	5	0.25	1	50

3.4.2 Proposed Circuitry

A specific circuitry has been developed to achieve higher attenuation. As shown in Fig. 3.11, the series inductors L_1, L_2, L_3 and L_4 block HF interference, while the capacitors C_1 and C_2 bypass noise currents. R_1 and R_2 are connected parallel to L_1 and L_2 respectively to reduce the Q-factor around 1.5 kHz. R_3 discharges C_1 and

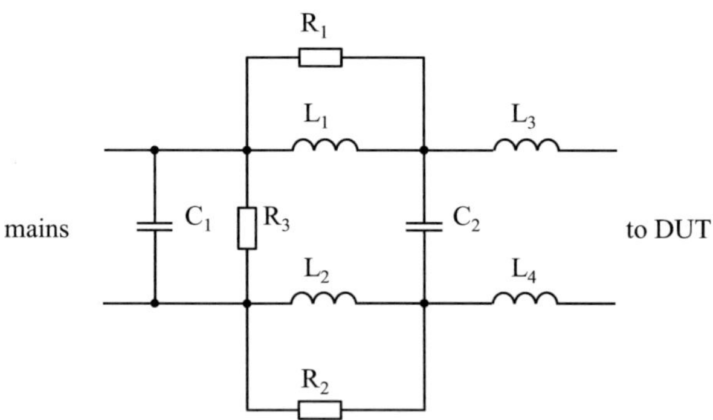

Fig. 3.11 Schematic diagram of the proposed LISN. The component values can have different combinations.

C_2 after the disconnection of the LISN. Table 3.2 shows an example of the parameter combination. This combination has lower capacitance values than the commercial LISN. Therefore, low-cost film or chip capacitors that are suitable for higher frequencies can be used in this circuit.

Table 3.2 Components of Proposed LISN

L_1 (mH)	L_2 (mH)	L_3 (mH)	L_4 (mH)	C_1 (μF)	R_1 (Ω)	C_2 (μF)	R_2 (Ω)	R_3 (MΩ)
2.5	2.5	2.5	2.5	1	330	1	330	1.5

The proposed LISN has higher attenuation in both directions, as shown in Fig. 3.12. For the direction from mains to DUT, the attenuation exceeds EN-1 by more than 20 dB. For the other direction, the difference is even larger.

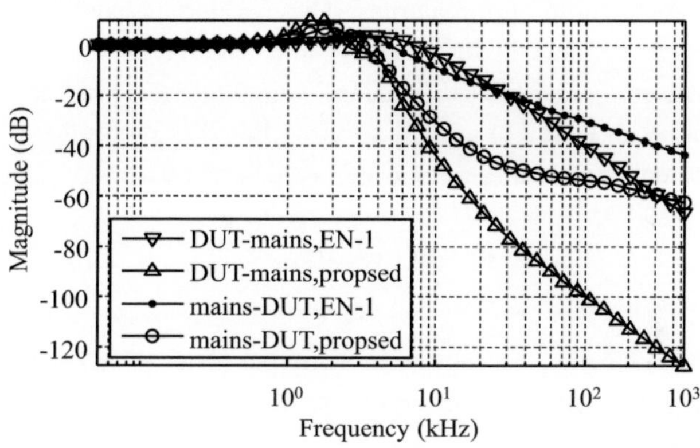

Fig. 3.12 Magnitude responses of different LISNs: mains to DUT.

The impedance seen by the DUT is mostly determined by L_3, L_4 and C_2. The frequency dependency is illustrated in Fig. 3.13. Obviously the absolute value increases with frequency above 1.5 kHz. It reaches 750 Ω and 12.6 kΩ at 30 kHz and 500 kHz respectively.

3.4.3 Toroid Inductor

The inductor in the specific LISN circuit can be implemented in different ways. The most commonly used inductors for high frequencies are toroid inductors with coils wound on a donut-shaped magnetic core, as shown in Fig. 3.14. Ferrite and

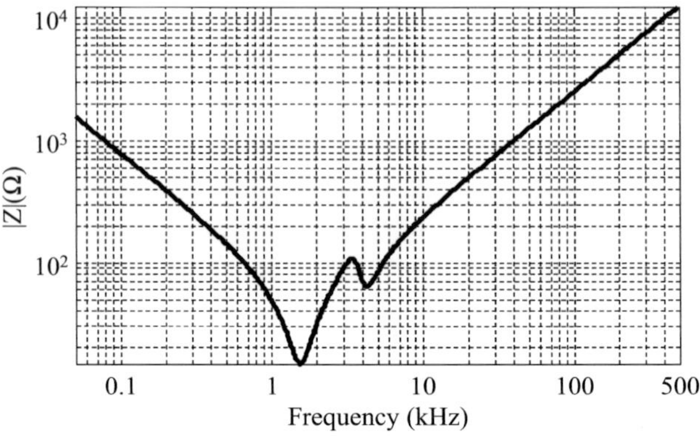

Fig. 3.13 Impedance seen by a DUT.

iron powder are favorite core materials. These materials have commonly large permeability values. Thus the toroid coils can reach high inductance with a small number N of turns. According to Ampere's circuital law, the line integral of a magnetic field B in tesla (T) around a closed line c is proportional to the total amount of current passing through the area s enclosed by c

$$\oint_c B \cdot dc = \mu \cdot \int_s J \cdot ds, \tag{3.3}$$

where μ is the magnetic permeability in henry per meter (H/m) and J denotes the current density within S. For the toroid coil illustrated in Fig. 3.14, this relation can be expressed by

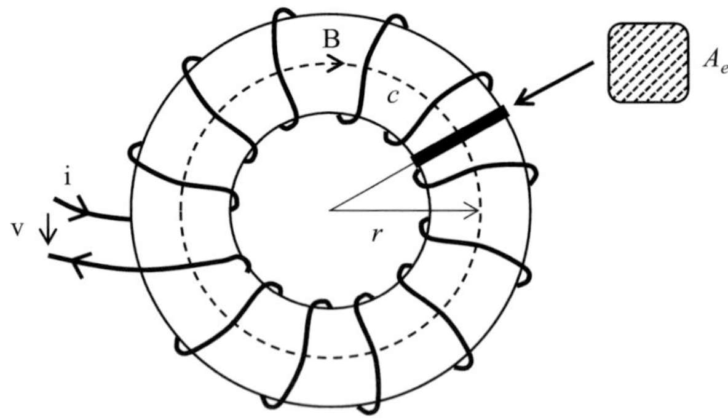

Fig. 3.14 Example of a toroid inductor.

$$B \cdot l_e = \mu_0 \cdot \mu_r \cdot N \cdot i, \tag{3.4}$$

where l_e denotes the effective length of the flux path c. μ_r is called relative permeability of the material. μ_0 is the permeability of free space, also called the magnetic constant. It has the value of $4\pi \cdot 10^{-7}$ H/m. The voltage induced by the changing magnetic field is obtained by applying Faraday's law

$$v = -N \cdot \frac{dB \cdot A_e}{dt}, \tag{3.5}$$

where A_e is the effective cross-sectional area of the core. The relation between the current flowing through the winding and the magnetic field strength H in ampere per meter (A/m) is

$$H = \frac{N \cdot i}{l_e}. \tag{3.6}$$

Although the ferro- and ferrimagnetic materials can provide large inductance, they suffer from magnetic saturation. These materials are composed of microscopic magnetic domains with randomly aligned directions. An external magnetic field H_E aligns the domains in parallel with it. The number of aligned domains depends on the strength of H_E. As shown in Fig. 3.15, magnetic saturation occurs when all the domains are aligned. After that, further increase of the applied mag-

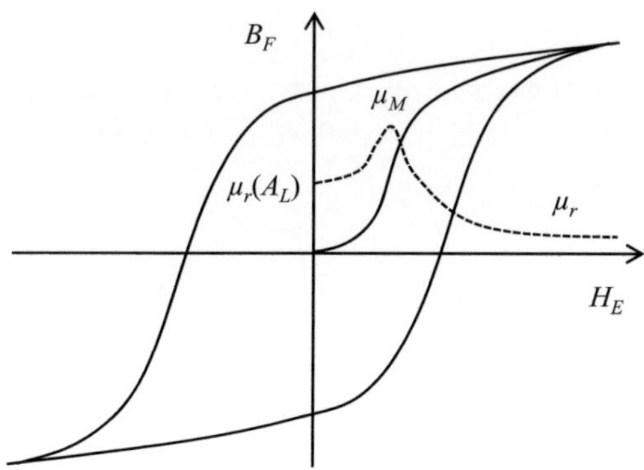

Fig. 3.15 Relations between H_E, B_F, and μ_r.

49

netic field can no longer change the internal status of the magnetic domains. As the material approaches the magnetic saturation, the flux density B_F approaches its maximum value. The relative permeability exhibits nonlinear behavior, i.e. it grows with increased applied magnetic field, reaches its maximum value μ_M and then declines toward one in the saturation state.

If the permeability is not a constant and is dependent on the applied external field, the differential permeability μ_{diff}, defined by [KAI04]

$$\mu_{diff} = \frac{1}{\mu_0} \cdot \frac{dB}{dH},$$

(3.7)

is often deployed. In the following the definition of differential permeability is used without further indication in the case of an alternate-current excitation, and it is also denoted by μ_r if no confusion occurs. The inductance L, defined as the flux change $d\Psi$ with respect to the current change di, given by

$$L = \frac{d(N \cdot A_e \cdot B)}{di} = \frac{\mu_0 \cdot \mu_r \cdot N^2 \cdot A_e}{l_e} \propto \frac{dB}{dH},$$

(3.8)

is also proportional to the slope of the B-H curve [KAI04]. The relation between the induced voltage v and L is

$$v = L \cdot \frac{di}{dt}.$$

(3.9)

Manufacturers of core materials often provide a material-dependent parameter called inductance factor A_L which is defined as the ratio of the inductance of a coil to the square of the number of turns:

$$A_L = \frac{L}{N^2}.$$

(3.10)

The A_L value provides a convenient way to calculate the inductance. However, it must be noted that the A_L values given by the manufacturers are usually measured at very low field strength (about 10 Gauss or 1 millitesla) [MAG88]. At very low field strengths (B is lower than 1% of the saturation flux density), the B-H curve is approximately linear [KAI04]. The permeability and thus the inductance factor could be different if the core is exposed to larger external magnetic fields. In our application, the core saturation must be avoided. It is important to find out the maximum current which will not cause core saturation. For the calculation, the inductance factor is assumed to be the same as the value provided by manufacturers. According to (3.4), (3.8) and (3.10) the maximum current i_{SAT} is

$$i_{SAT} = \frac{B_{SAT} \cdot A_e}{N \cdot A_L}. \tag{3.11}$$

B_{SAT} is the magnetic flux density above which the core is saturated. It is also provided by manufacturers. Table 3.3 shows the parameters of the ferrite core which is investigated here.

Table 3.3 Parameters of Toroid Inductor

N	A_L (nH/N^2)	A_e (mm^2)	L (mH)	B_{SAT} (mT)	i_{SAT} (A)
34	2150	51.26	2.48	500	0.35

3.4.4 Unwanted Impulsive Noise

Depending on load conditions (impedance of the connected DUT), some periodical large impulsive voltages may appear at the LISN output. This impulse sequence has a repetition rate of twice the mains frequency. It aligns itself with the zero-crossings of the mains current. Loads with large capacitance can excite the impulsive noises significantly. An increased number of loads can also give rise to and amplify the impulses. The spectral characteristics of the individual impulse are also dependent on the load conditions. Sometimes they contain spectral components located in the same frequency range as the transmitted signal. In this case the test result will be affected by these artificial impulses. They must be eliminated so that accurate test results can be obtained.

To investigate the origin of the impulsive noise, a measurement setup is built and its block diagram is shown in Fig. 3.16. For simplification the resistors R_1 and R_2 shown in Fig. 3.11 are not illustrated here. A test signal u_T is superimposed on the mains voltage u_M. It is a sinusoidal waveform with 33 kHz. The currents flowing through L_1 and L_2, denoted by i_{L1} and i_{L2}, and the voltage across L_1, denoted by u_{L1} are measured at the same time. High-frequency components of the voltages on the mains' and the DUT's sides, denoted by u_{HP1} and u_{HP2} respectively, are picked up using two identical high pass filters.

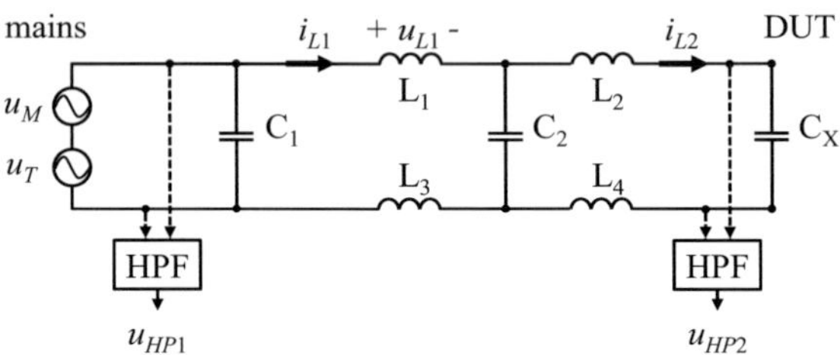

Fig. 3.16 Block diagram of the measurement setup.

Fig. 3.17 shows the measurement result. As can be seen in plot (a), the test signal is visible with relatively constant amplitude. In its short-time Fourier transform (STFT), the spectral component at around 60 kHz belongs to a narrowband interferer from the mains network. Periodical impulsive noise is synchronous with the zero-crossings of the mains current. Their wideband spectra exceed 200 kHz. The waveform of i_{L1} is shown in plot (c). The impulsive noise is superimposed on the mains current too. The amplitude of i_{L1} is slightly larger than that of i_{L2} since a part of i_{L1} flows through C_2. Plot (d) shows the waveform of u_{HP2}. Obviously, the impulses have larger averaged magnitude than those in u_{HP1}. Plot (e) shows the STFT of u_{HP2}. The impulsive noise is visible. Parts of its spectrum have been attenuated and its spectral components above 100 kHz cannot be observed any more. However, its low-frequency parts seem to be amplified instead of attenuated.

The power spectral density (PSD) of the test signal shows an obvious periodicity. The test signal seems to be switched on and off periodically with a repetition rate of 100 Hz. The "ON" state follows the impulses and lasts for about 6.5 ms. It corresponds to the interval in which the magnitude of the mains current exceeds 0.15 A. The PSD of the test signal is not constant within the "ON" state. It reaches the maximum in the center. It is reasonable to associate the periodicity of the test signal with a repeated saturation of the ferrite core. As mentioned in 3.4.3, the core saturation leads to a drastic reduction of the permeability. As a result, the attenuation of high frequency components decreases too. However, the core saturation does not mean a total disappearance of the inductance. Therefore, the LISN still blocks the interferer at 60 kHz. That's why only the 33 kHz test signal is still visible in the last plot in Fig. 3.17. The reason for the periodical impulses is more complicated. It's worth mentioning that these impulses can still be observed even if the mains voltage at the LISN input has been filtered appropriately. Therefore,

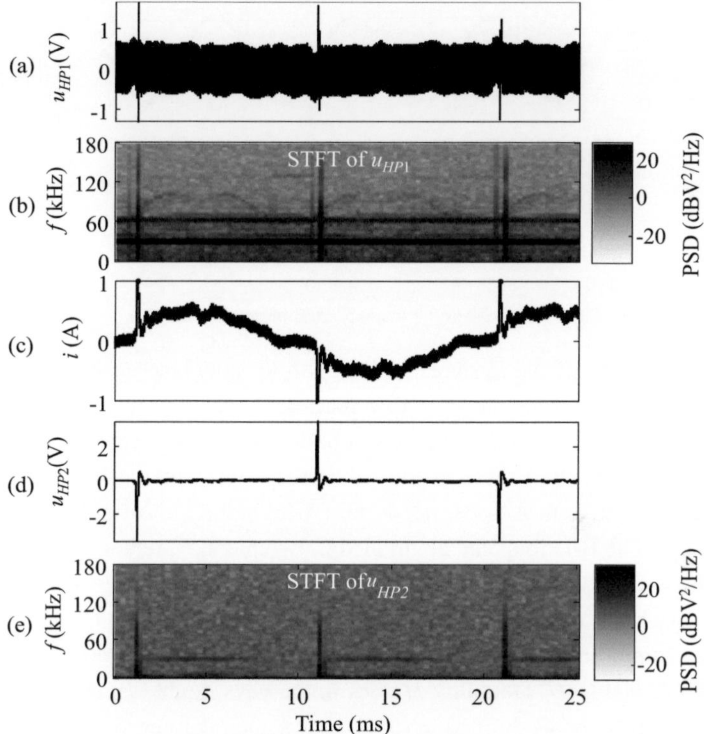

Fig. 3.17 Observation of saturated ferrite cores.

the hypothesis that the impulses are originating from the mains, and are less attenuated by the LISN due to core saturation, turns out to be incorrect. Meanwhile, it has been observed that these impulses appear when either active DUTs or passive RC components are connected to the LISN. Therefore, it can be excluded that these impulses are generated by the DUTs. All clues point to a load-dependent non-linear behavior of the LISN.

Fig. 3.18 shows the measured induced voltage u_{L1} and the current i_{L1} that flows through the inductor. i_{L1} still maintains the sinusoidal waveform. u_{L1} has the greatest peaks when i_{L1} crosses zero.

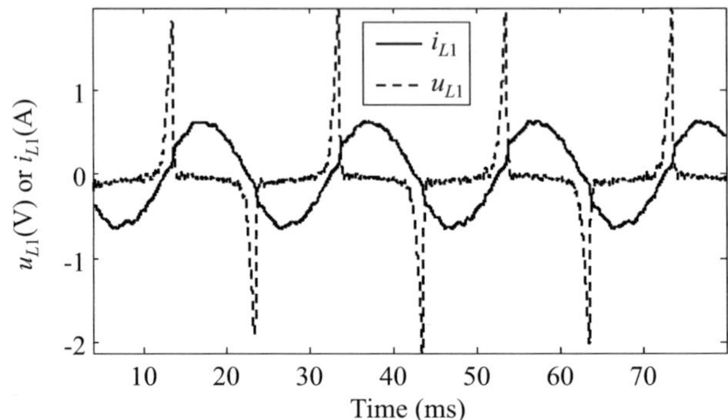

Fig. 3.18 Measured current i_{L1} and induced voltage u_{L1}.

An investigation of the hysteresis of the toroid inductance is necessary to explore its nonlinear behavior. For this purpose, a single inductor is involved. Measurement of the dynamic hysteresis loop is carried out according to [IEEE91]. Fig. 3.19 illustrates the B-H loop and the corresponding permeability curves. For convenience, both H in Ampere per meter (A/m) and i_{L1} in Ampere are given for the abscissa. The B-H loop can be divided in two parts: B_1 refers to the process in which the excitation current i_L (also the external field strength H) increases from the negative peak to its positive maximum. The ferrite core is brought from its negative saturation -B_{sat} to the positive saturation B_{sat}. As expected, B_1 is not linearly proportional to H. Doubling H does not always double B_1 correspondingly. The permeability μ_{r1}, as defined in (3.7), shows a similar curve as the one shown in Fig. 3.15. However, the permeability in Fig. 3.15 is only for an initial magnetization process, while Fig. 3.19 shows the change of the permeability in an alternating magnetic field. μ_{r1} is extremely low when the ferrite core is in the negative saturation state. It increases slowly with increasing field strength and reaches its maximum when B_1 passes through the abscissa.

After that it decreases towards 1, when the ferrite core falls into saturation. If the excitation changes its direction from B_{sat} to -B_{sat}, the hysteresis loop traces B_2. The process is similar, except that μ_{r2} is now proportional to the negative deriva-

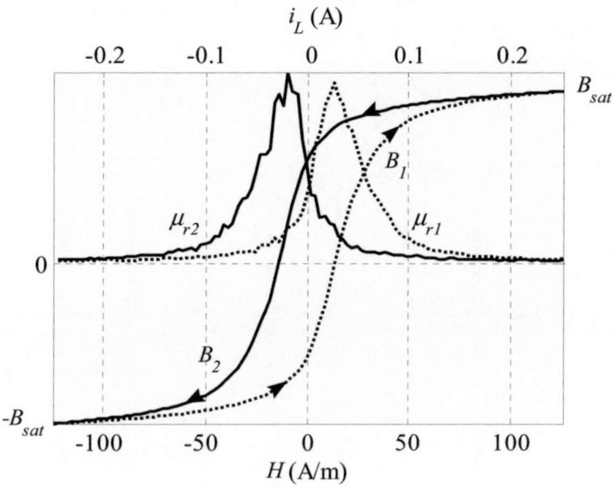

Fig. 3.19 Measured traces of H, B, and μ_r.

tive of B_2. The peaks of μ_{r1} and μ_{r2} are almost symmetrical with respect to the y-axis. Since the inductance is proportional to the permeability, it is also dependent on the excitation current and exhibits a maximum in the vicinity of the zero-crossing points of the current. If the excitation current changes periodically with a sinusoidal waveform, the permeability and the inductance also change periodically. Recall the similar current-dependent feature of impulsive noise; it might be reasonable to associate the noise source with the alteration of the permeability.

In order to verify the assumption, a simulation is made for the circuitry shown in Fig. 3.16. Capacitor C_2 is discarded for simplicity. A sinusoidal waveform is generated for the mains voltage u_M and is shown in plot (a) in Fig. 3.20. The amplitude and the frequency of u_M are 325 V and 50 Hz respectively. Since this simulation only handles the impulsive noise, the test signal u_M is replaced by a serial passive component R_M. It can be treated as the equivalent output resistance of the mains. The current-dependent inductance is considered for the inductors L_1 through L_4. The capacitors C_1 and C_X are assumed to be "ideal" components, so that the analysis can focus on the inductors. C_1 has a constant value of 1 µF, the same as in the LISN. 3 µF is assigned to C_X since it is the typical value of loadings of the LISN when the impulsive noise is observed. The simulation is carried out in the time domain. Three fundamental relations are used to describe the circuitry.

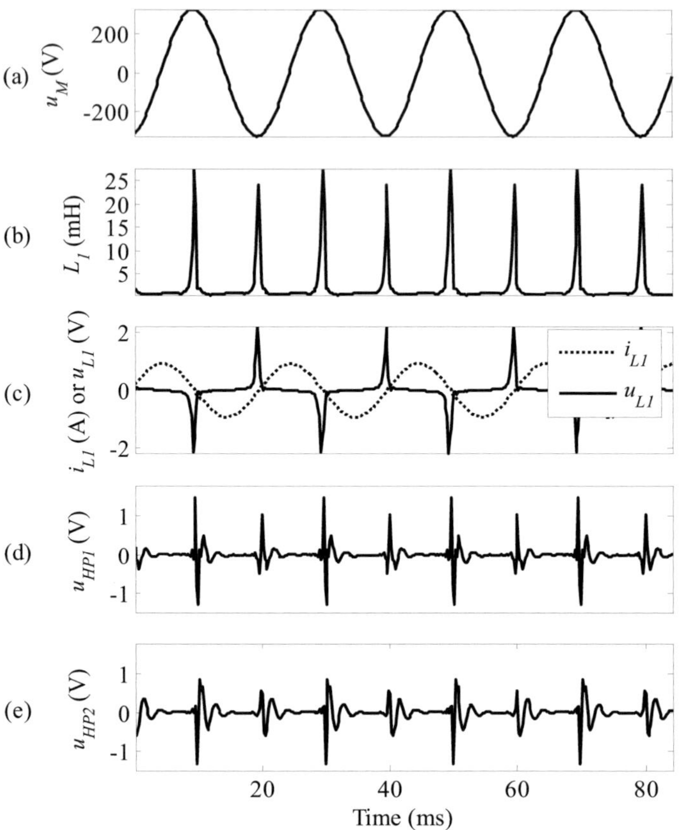

Fig. 3.20 Simulated origin of impulses.

$$u_{C_1} = 4 \cdot L_1\left(i_{L_1}\right) \cdot \frac{di_{L_1}}{dt} + u_{C_X}, \tag{3.12}$$

where u_{C1} and u_{CX} denote the voltages cross capacitors C_1 and C_X respectively. The inductance L_1 depends on i_{L1} and the relation is given by the permeability curves μ_{r1} and μ_{r2} as well as (3.8). The mains voltage consists of the voltages across R_M and C_1, i.e.

$$u_M = R_M \cdot \left(i_{L_1} + C_1 \cdot \frac{du_{C_1}}{dt} \right) + u_{C_1},$$ (3.13)

and the current through L_1 is the same as that flowing through C_X., i.e.

$$i_{L_1} = C_X \cdot \frac{du_{C_X}}{dt}.$$ (3.14)

L_1, i_{L1} and u_{L1} are calculated and shown in plot (b) and plot (c). The inductance changes periodically. In addition, the change is synchronous with the absolute maximum of u_M or the zero-crossings of i_{L1}. The inductance varies between 1 and 26 mH. i_{L1} still keeps the sinusoidal waveform, except some distortions next to the zero-crossing points. u_{L1} reaches its positive and negative maximum at the rising and falling edges of i_{L1} near the zero-crossings. The simulated i_{L1} and u_{L1} agree well with the measured counterparts shown in Fig. 3.18. Intuitively, the peaks of u_{L1} are given by a multiplication of the inductance maximum with the largest derivation of i_{L1}. Plot (d) and (e) show u_{HP1} and u_{HP2} obtained by filtering u_{C1} and u_{CX} respectively. The simulation has shown that the inductance of the ferrite core inductor varies with the current flowing through the windings. This current-dependency comes from the non-linear nature of the ferrite core materials. The periodic mains voltage drives the inductance up and down in a cyclic manner. The peaks can have values more than ten times larger than the initial value. They appear next to the zero-crossings of the mains current where the slope of the current reaches its maximum. This synchronized phenomenon results in large voltage peaks cross each inductor. They are the origins of periodical impulsive noise. Depending on the load condition and the equivalent access impedance of the mains, the magnitudes of these impulses at the mains' and the load's sides may be different. The impulses at the load's side can even be stronger than at the mains' side under certain conditions.

3.4.5 Air Core Coil

An inductor can also reach a specified inductance without any magnetic core. The turns can be wound on nonmagnetic materials, or even have only air inside them. Now permeability and inductance are independent of the current. The B-H curve is linear, i.e. non-linearity due to material saturation will no longer appear. Thus, an air filled coil allows arbitrary large currents, in contrast to a magnetic core inductor. Nevertheless, absence of a high permeability core results in low inductance and large stray fields. A high number of turns is necessary to achieve a considerable inductance in the range of mH. A multilayer air core design is a common implementation of an inductor with many turns. As shown in Fig. 3.21, the coil is wound on a plastic coil bobbin.

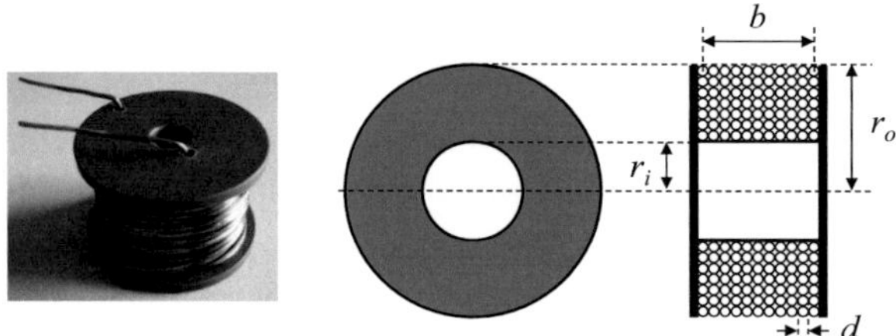

Fig. 3.21 Multilayer air core inductor.

Assume the winding has a diameter d. The coil has inside radius r_i, length b and outside radius r_o. Its inductance L can be approximated by [WHE28]

$$L \approx \frac{7.9 \times 10^{-6} \cdot \left(r_i + r_o\right)^2 \cdot N^2}{13 \cdot r_o - 7 \cdot r_i + 9 \cdot b} \tag{3.15}$$

In practice, it is desirable to calculate the number of turns N to achieve a specific inductance L for a given coil bobbin (given r_i and b) and a certain type of wire (known diameter d). The number of wires on each layer is

$$N_0 = \left\lfloor \frac{b}{d} \right\rfloor. \tag{3.16}$$

The outer radius r_o is a function of N, i.e.

$$r_o = r_i + \frac{N}{N_0} \cdot d. \tag{3.17}$$

Inserting (3.17) into (3.15) gives

$$\frac{31.6\times10^{-6}}{L}\cdot\left(r_i+\frac{N\cdot d^2}{2b}\right)^2\cdot N^2-\frac{13d^2}{b}\cdot N=6r_i+9b. \tag{3.18}$$

The symbolic expression for N is quite complicated. Instead, it is much easier to split this equation in two functions of N as follows

$$y_1(N)=\frac{31.6\times10^{-6}}{L}\cdot N^2\cdot\left(r_i+\frac{d^2}{2b}\cdot N\right)^2, \tag{3.19}$$

and

$$y_2(N)=6r_i+9b+\frac{13d^2}{b}\cdot N. \tag{3.20}$$

If both functions are plotted together with N as abscissa and the function values as ordinate, as shown in Fig. 3.22, both curves cross at one point in the first section where N, y_1 and y_2 are positive values. The projection of the intersecting point onto the abscissa corresponds to the expected number of turns. In this example, the expected inductance value is 2.5mH, and the other dimensional pa-

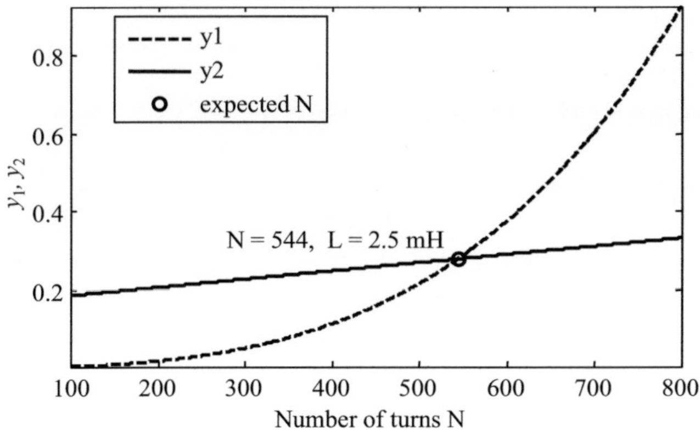

Fig. 3.22 Curves for y_1(N) and y_2(N), N>0.

rameters are listed in Table 3.4. Totally 544 turns are needed to achieve the expected inductance. The designed inductor is examined with a vector network analyzer. The inductance and the capacitance in the frequency range from 30 kHz to 1 MHz are shown in Fig. 3.23. Obviously, the air core exhibits a resonance between 700 and 800 kHz. The inductance is relatively flat below the resonant frequency and has a value of about 2.48 mH. Above that frequency the coil is capacitive and the parasitic capacitance is about 14 pF at 1MHz. Parameters of the equivalent circuit, such as the winding resistance R_S and the stray capacitance C_P can be seen in Table 3.4. Despite the resonance, the inductor is well applicable in the frequency range of interest.

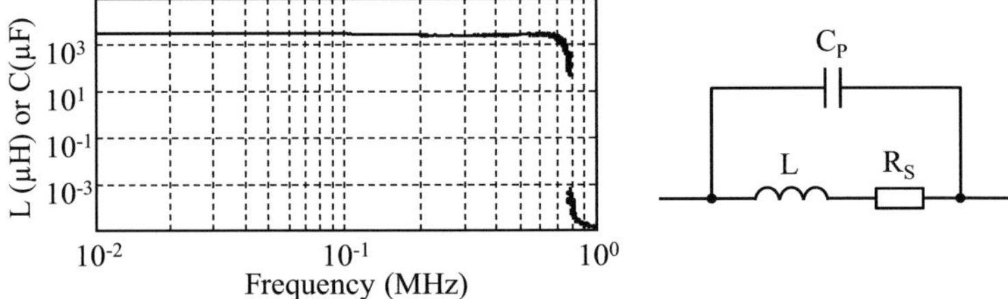

Fig. 3.23 Measured inductance and capacitance of the air core inductor.

Table 3.4 Parameters of an Air Core Inductor

L (mH)	d (mm)	b (mm)	r_i (mm)	N	C_P (pF)	R_S (Ω)	f_R (kHz)
2.48	0.5	16	4	544	16	2.9	788

3.5 Uninterruptible Power Supply (UPS)

A high-quality and noise-free mains voltage is an important prerequisite for the testbed. Although the aforementioned LISN circuitry can filter out large portions of HF noise, it has low attenuation in low frequency range, especially below 10 kHz. Large transients can still penetrate the LISN and disturb the transmitted signal. In addition, amplitude fluctuation and frequency instability of the mains voltage could also introduce uncertainties into the test process. Therefore, additional effort is needed to ensure a reliable source of power supply. Consequently, a double conversion UPS is used here for this purpose. The UPS is a device which provides a stable and high quality power supply in case of failure and interruption of the mains power. Fig. 3.24 shows a simplified block diagram of a double conversion UPS. A rectifier and an inverter are connected in series between the input and the output. A battery is connected to the output of the rectifier. During normal operation, the mains voltage is converted to direct current (DC) which charges the battery and drives the inverter. The inverter converts DC to AC and delivers a well-controlled and regulated supply voltage to the loads. If the mains voltage fails or runs out of the specified tolerance range, the battery - instead of the rectifier - begins to drive the inverter. The power flow always undergoes an AC-DC and a DC-AC conversion. This explains why this kind of system is called double conversion UPS.

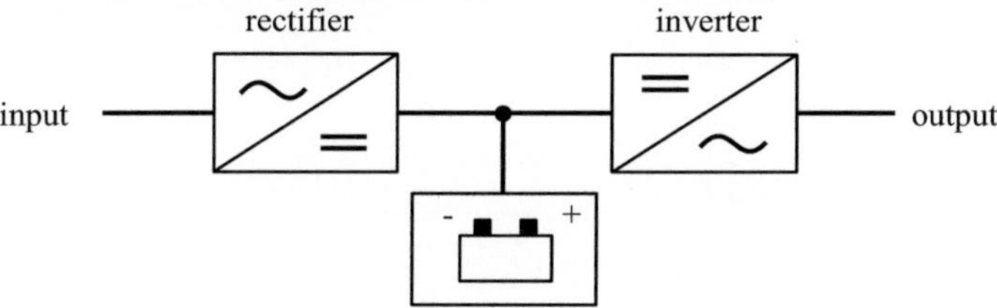

Fig. 3.24 Simplified block diagram of a double conversion UPS [FUR02].

Many modern inverters are operating with a pulse-width modulation (PWM) control scheme. Spectral components at the switching frequency and harmonics introduce narrowband interference. Fig. 3.25 (a) shows a waveform measured at the output of the UPS. High frequency ripples shown in plot (b) is superimposed on the synthesized mains voltage. Fig. 3.26 illustrates the STFT of the waveform. Narrowband interference at 20 kHz and its harmonic at 40 kHz are visible.

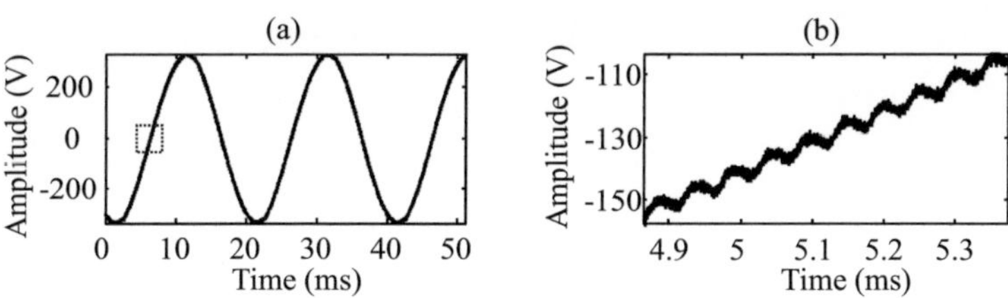

Fig. 3.25 Waveform at the UPS output.

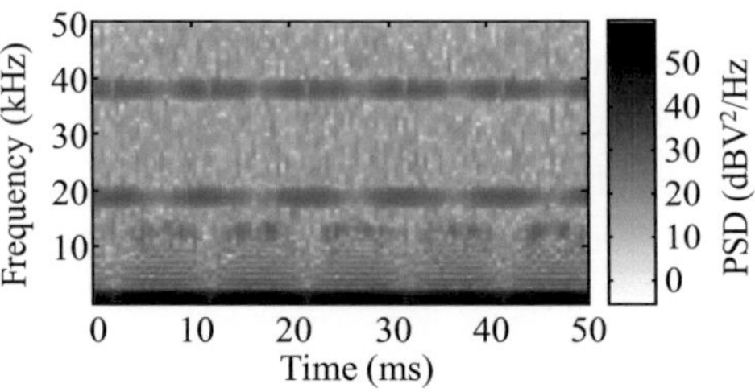

Fig. 3.26 STFT of the UPS output.

Fortunately the spectral components are stationary and fall into the range where the LISN has high attenuation. Therefore, they can be filtered out without further effort. The original and the filtered spectrum are shown in Fig. 3.27.

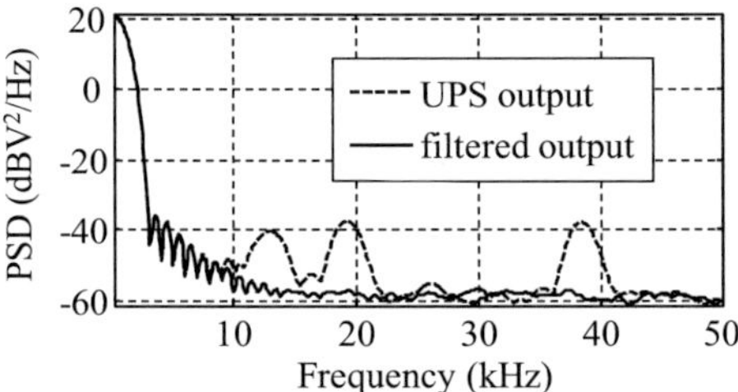

Fig. 3.27 Original and filtered UPS noise.

4 Channel Transfer Function

The CTF is a fundamental characteristic of a data transmission medium. A complex-valued CTF $H(f)$ can be defined as

$$H(f) = |H(f)| \cdot e^{j\varphi(f)}, \tag{4.1}$$

where $|H(f)|$ and $\varphi(f)$ are the amplitude and phase responses respectively. A non-constant amplitude response indicates a frequency selective attenuation. A non-linear phase response causes a phase distortion. Group delay $\tau(f)$ can be obtained by the derivative of $\varphi(f)$ with respect to f

$$\tau(f) = -\frac{1}{2\pi} \cdot \frac{d\varphi(f)}{df}. \tag{4.2}$$

It is a convenient measure of the linearity of a channel's phase response. A constant $\tau(f)$ indicates a linear phase course, while any variation of $\tau(f)$ over frequency results in a phase distortion of transmitted signals [FER2010].

4.1 Chanel Transfer Function (CTF) for NB-PLC

Many measurements have revealed that the signal propagation in narrowband channels depends on the frequency and that there is larger attenuation in the low frequency region. Typical PLC channels are characterized by high dynamics with respect to their amplitude response and partly high attenuation, within both intra-building and access domains [CHA86] [BAU06]. Narrowband frequency-selective fading can occur depending on the load. Deep notches and heavy growth of attenuation have been observed for environments described in [DOS00]. In contrast to wideband PLC channels where the course of the frequency response is dominated by multipath propagation, the frequency selectivity of NB-PLC channels is mainly caused by appliances (including circuit breakers) connected to the lines. Besides, due to the time-variant nature of the loads, the transfer functions also change with time [CHA86] [DOS00] [FER10]. It has been reported in [SUG08] that time-variance is not visible when no electric appliances are connected to outlets. In addition, notches are seldom in the magnitude response and the propagation attenuation exhibits smoothed traces. When electric appliances are connected to outlets, time variation can be found, however, solely in particular frequency ranges. In case of phase distortions, the communication quality of most phase modulation techniques may degrade. For example, the time domain DPSK may be disturbed by time-variant phase distortions, and the frequency domain DPSK used by PRIME may suffer from phase distortion. Matched filter-based modulation - for example DCSK- and preamble-based synchronization can also be impaired if the correlation between the received signal and the template is reduced by phase distortions.

Generally, for NB-PLC most effort was focused on the investigation of the amplitude response, while the phase response has either been ignored or idealized as time-invariant. The earliest investigation of phase shift for NB-PLC has been reported in [HOO98], where a sine wave was transmitted over 250 meters of residential power circuit (RPC) and made visible on an oscilloscope after filtering away the channel noise. In parallel, the same sine wave was transmitted over a twisted pair connection and displayed on the second oscilloscope channel. From the amount of drift of the signal transmitted over the RPC channel with respect to the reference signal, the amount of time variance of the RPC-induced phase shift could then be determined. Variations with respect to the reference were never larger than about 10 degrees. However, due to increased complexity of LV powerline networks, significant time-variance and non-linear phase behavior may occur. Periodic sharp phase changes synchronized with mains cycle have been observed in [SUG09]. Thus, also the quality of popular differential phase modulation methods may be severely degraded if no proper compensation of the cyclic changes is made. In order to characterize a NB-PLC channel completely, it is necessary to investigate both the attenuation and the phase distortion.

4.2 Measuring Complex-Valued CTF

4.2.1 Overview of Measurement Approaches

Since the magnitude measurement is relatively simple and straightforward, it is worth overviewing the state of art with focus on group delay measurements. At the same time, possibilities of integrating the magnitude measurement into the measurement methodology are also to be considered. Group delay is often measured by using a vector network analyzer (VNA). A simple and common procedure obtains the phase response during a S-parameter measurement and approximates the group delay by calculating the phase differential quotient of S21

$$\tau \approx -\frac{1}{360°} \cdot \frac{\Delta\varphi}{\Delta f}. \tag{4.3}$$

The integrated transmitter and receiver in the VNA share the same trigger source. Therefore, any phase shift $\Delta\varphi$ between the transmitted and received signals can be measured with very high accuracy. If the phase shift within Δf does not suffer from severe non-linearity, this method delivers satisfying results. A properly selected Δf and calibration of the VNA can further improve the resolution, as well as the accuracy [OST97]. This method is mostly used for investigating the phase distortion of non-frequency-translating devices such as amplifiers and filters. However, an accurate determination of absolute phase courses becomes challenging as the frequency increases.

To reduce the effort at high frequencies and to make the measurement independent of local oscillators, modulation-based methods have been used as low-cost but accurate alternatives for the common VNA approach. This category of methods modulates a sinusoidal waveform with a single frequency at which the group delay is to be investigated. The modulated signal is then transmitted through the DUT, and the phase shift between envelopes of the input and output signals is measured. The expected group delay is proportional to the envelope phase shift

$$\tau \approx -\frac{1}{360°} \cdot \frac{\Delta\varphi}{f_m}, \tag{4.4}$$

where $\Delta\varphi$ is the envelope phase shift and f_m is the modulation frequency. The modulation techniques include amplitude modulation (AM) [HP68] [BAD85] [KER87], frequency modulation (FM) [WOO99] [WOO05], or phase modulation (PM). The main advantage of the modulation-based measurements is that the phase measurement is performed at a much lower frequency. This can be done precisely using any low-cost phase detection device. The envelope can also be converted to digital values with low speed ADCs and its phase shift can then be estimated using modern digital processing algorithms. Nevertheless, the disadvantage, as mentioned in [WOO05], is the difficulty of predicting and controlling measurement errors. To reduce measurement uncertainties and to control the error sources, a new method, called two-tone method, has been developed. Instead of modulating a single frequency signal this approach needs two frequencies, f_1 and f_2, for the test signal. The relative phase difference between these two frequencies is measured at both the input ($\Delta\theta_o$) and the output ($\Delta\theta_i$) of the DUT. The desired group delay is estimated by

$$\tau \approx -\frac{1}{360°} \cdot \frac{\Delta\theta_o - \Delta\theta_i}{f_2 - f_1}. \tag{4.5}$$

Despite all differences, the above-mentioned sounding techniques have two features in common: they use narrowband sounding signals and ideal synchronization between the transmitter and the receiver. In the real-world measurements of powerline channels, the transmitter and the receiver are usually placed at different locations. It is quite difficult to carry out the measurement without an accurate global time reference. To overcome the synchronization problem, [VAN90] has deployed periodical broadband signals to obtain the channel group delay. Totally N measurements of the same duration are made with random delays $\Delta\tau$ among them. Each measurement lasts for multiple periods of the sounding signal. The phase difference $\Delta\theta$ between two adjacent frequencies is calculated for each measurement:

$$\Delta\theta_m(i) = \theta_m(i) - \theta_m(i-1) + \left[\omega(i) - \omega(i-1)\right] \cdot \Delta\tau_m, \tag{4.6}$$

where m is the measurement index ($m = 1, 2, ..., N$), and i denotes the frequency index. After that, the frequency-dependent phase difference is averaged over all measurements, according to

$$\Delta\widehat{\theta}(i) = \frac{1}{N} \cdot \sum_{m=1}^{N} \Delta\theta_m(i) = \Delta\theta(i) + \left[\omega(i) - \omega(i-1)\right] \cdot \frac{1}{N} \cdot \sum_{m=1}^{N} \Delta\tau_m. \tag{4.7}$$

$\Delta\theta(i)$ denotes the expected phase difference. The influence of a random delay $\Delta\tau$ on the phase difference is reduced to a random constant which is independent of frequency. By averaging all measurements, these constants are added together, but they do not change the shape of the phase difference. If the random delay is an unbiased random process, the second term in (4.7) approaches 0 and the estimated phase difference approaches the expected value. Candidates for an excitation signal can be periodic chirps, optimized multi-tone sine waves, or maximum length binary signals (MLBS). The author in [VAN90] has chosen MLBS because of the low cost for implementation. Although the shape of the group delay can be recognized in the work of the authors, the results still suffer from some uncertainty.

A multi-carrier sounding technique has been deployed in [SUG08]. The sounding signal is a linear sum of multiple sinusoidal waveforms. As a result, its spectrum consists of multiple equidistant peaks, one for each sinusoidal waveform. DFT is performed to the distorted and attenuated signal in order to obtain the signal attenuation. This method is efficient and fast since the whole frequency domain of interest can be investigated simultaneously. The frequency selective and time-variant features can also be observed conveniently. However, since the transmitted power is distributed over multiple frequencies, the SNR at each frequency is reduced. Therefore the measurement accuracy degrades if the SNR is low.

Due to the uniqueness of NB-PLC channels, the sounding signal shall deliver as much power as possible at the frequencies of interest, so that the received signal exhibits high SNR values. The approach shall be able to determine as many parameters as possible from a single data acquisition. Parameters of interest include amplitude response, phase distortion, noise scenarios and their time-variant behavior. It is also desirable to investigate harmonic distortions caused by non-linear electronic components. In addition, the method must allow the receiver to perform simple automatic gain control (AGC) algorithms without disturbing the amplitude response and noise power measurements. Under consideration of all above factors, the baseband double-frequency method is chosen and extended for the measurements involved throughout this thesis.

4.2.2 Proposed Measurement Platform

Fig. 4.1 shows a measurement platform for powerline channel characterization. A global positioning system (GPS) receiver provides a global time reference with an accuracy of 0.1 µs. A FPGA controls data acquisition, time stamping and signal generation. A hard disk drive (HDD) is used to store signals and timestamps for long-term measurements. An AFE - equipped with a 12-bit ADC and a 14-bit DAC - converts analog signals to digital values for signal acquisition and digital values to analog signals for the signal generation respectively. The signal generation path is followed by a line driver whose output can be switched to high impedance when the platform is working as a receiver. Configured as a transmitter, it can inject sounding signals into the mains with a maximum of 16 volts peak-to-peak (V_{pp}) and an average current of more than 1.5 amperes (A). Separate coupling components C_T and C_R are used for the transmitting and receiving paths respectively. C_T is optimized for the frequency range up to 500 kHz in such a way that its equivalent output impedance is lower than 0.5 Ω, and the signal transformer can carry large current without noticeable non-linear distortions. These measures guarantee high injection power, even if the access impedance is very low. On the contrary, C_R has large input impedance. Therefore, it will not influence the electrical characteristics of the measurement point. A zero-crossing detector provides information about the mains frequency. Two of these platforms are needed for a point-to-point measurement. During each measurement one platform is configured as a transmitter and the other as a receiver.

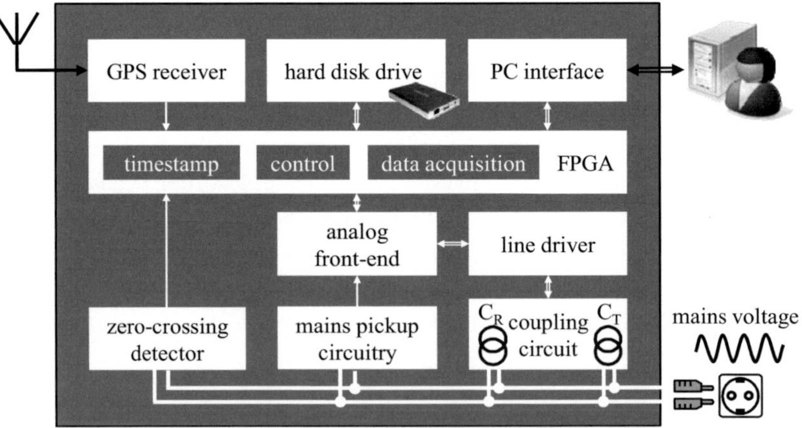

Fig. 4.1 Platform to measure complex-valued channel transfer function.

4.2.3 Discrete Double-Frequency Approach

The sounding signal $x(t)$ is a sum of two sinusoids with instantaneous phase $\theta(k)$ and the same amplitude A:

$$x_i(t) = \sum_{k=i,i+1} A(k) \cdot \sin[2\pi \cdot f(k) \cdot t + \theta(k)], \tag{4.8}$$

where $f(i)$ denotes the ith frequency component. The received signal $y(t)$ is

$$y_i(t) = \sum_{k=i,i+1} B(k) \cdot \sin[2\pi \cdot f(k) \cdot t + \theta(k) + \Delta\theta(k)], \tag{4.9}$$

where B and $\Delta\theta$ denote the attenuated amplitude and the phase shift respectively. The magnitude of attenuation can be obtained by

$$|H(i)| = \frac{B(i)}{A(i)}. \tag{4.10}$$

The group delay is approximated by

$$\tau(i) \approx \frac{-1}{360°} \cdot \frac{\Delta\theta(i+1) - \Delta\theta(i)}{\Delta f}. \tag{4.11}$$

The frequency difference, also called aperture is

$$\Delta f = f(i+1) - f(i). \tag{4.12}$$

The phase response $\varphi(i)$ can be approximated by

$$\varphi(i) = \theta(0) - 2\pi \cdot \sum_{k=1}^{i} \tau(k) \cdot \frac{f(k) - f(k-1)}{2}. \tag{4.13}$$

4.2.4 Measurement Setup

Two platforms - AMS1 and AMS2 - are deployed for a bidirectional point-to-point measurement. For each direction, one platform is configured as a transmitter, while the other is a receiver. The transmitter can record its own sounding signal during the transmitting interval. This feature is necessary for an accurate measurement since the mains access impedance can distort the sounding signal at the transmitter. With this recording-while-transmitting ability it is capable of recording the distorted transmitted signal and compensating any deviations later in the signal processing step. The bidirectional measurement is done in a time-multiplexing manner: time slots are assigned to both platforms. As shown in Fig. 4.2, a symbol is transferred from one point to the other in each slot. In the successive slot the direction changes. Each sounding symbol lasts for 800 ms, followed by a 200 ms pause. For either direction, the investigated frequency increases by one aperture in a new slot. Having reached the stop frequency, it returns to the start frequency and repeats the same procedure. The measurement covers the frequency range 20 - 500 kHz. The discrete frequency in kHz can be expressed as

$$f(i) = 19.5 + \Delta f \cdot i \tag{4.14}$$

where $i = 1, 2, \ldots, 961$. The symbol duration T_i is set to multiple mains cycles, so that any cyclic time-variance can be detected. The update rate of the signal generation is 10 MHz, and the sample rate of the data acquisition is 2 MHz.

4.2.5 Signal Processing

The acquired test symbols are processed offline using STFT. Each obtained sounding symbol $x(k)$ is divided into L overlapping segments. Each segment has

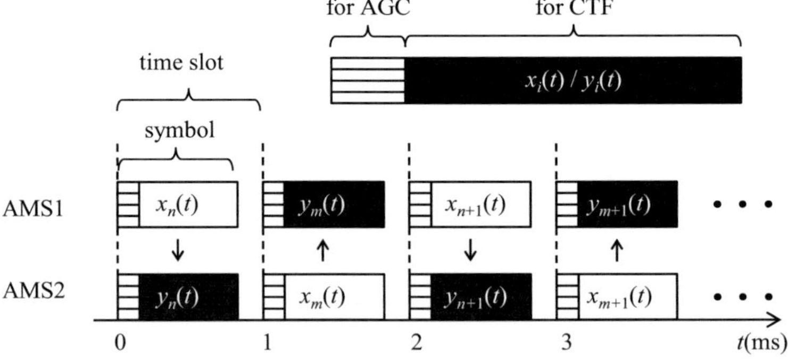

Fig. 4.2 Measurement time slots.

N_{SEG} sample values. Any two adjacent segments are overlapping by N_0 samples. Each individual segment is windowed and converted from the time to the frequency domain via DFT, according to

$$X(f,l) = \sum_{n=-\infty}^{\infty} x(n) \cdot w(n-l) \cdot e^{-j2\pi f \cdot n}, \tag{4.15}$$

where $X(f,l)$ and $w(n)$ denote the Fourier transform of the lth windowed segment and the window function respectively. $X(f,l)$ is complex-valued, carrying both magnitude and phase information. The magnitude is the absolute value of $X(f,l)$ and the instantaneous phase at a frequency f_i in the current segment is obtained by

$$\theta(f,l) = \Im\{\ln[X(f,l)]\}. \tag{4.16}$$

The amplitude response and the group delay can be obtained according to (4.10) and (4.11) by

$$|H(i,l)| = \frac{B(i,l)}{A(i,l)} \tag{4.17}$$

and

$$\tau(i,l) = \frac{-1}{2\pi} \cdot \frac{\Delta\theta(i+1,l) - \Delta\theta(i,l)}{\Delta f}. \tag{4.18}$$

4.2.6 Error Analysis

4.2.6.1 Principle Error

Theoretically, the group delay is defined as the negative derivative of a phase response with respect to frequency. In the proposed approach, the differential quotient versus aperture rather than the derivative is calculated. The measured value approaches the definition only if the aperture becomes infinitely small. However, the aperture can never be that small in reality, yielding a deviation of the measured value from the true value. Since this error originates by the measurement method, it can only be reduced but not be eliminated. This error is largely influenced by the aperture. A proper aperture is a compromise between accuracy and resolution. Generally speaking, the sensitivity of phase measurement is proportional to the aperture. A broader aperture could result in a larger phase shift, which can be measured with higher accuracy. However, if the two frequencies are too far away from each other, the details of group delay variations between both frequencies cannot be detected, and the resolution is reduced. The phase shift between two adjacent frequencies should not exceed 180°. Otherwise phase ambiguity occurs which makes it difficult to distinguish between a phase shift of φ_0 and $\varphi_0 + 2n\pi$ [OST97]. For the choice of an appropriate aperture one shall consider the desired accuracy, the group delay characteristics of the channel, and the frequency resolution of interest. In order to estimate the proper aperture, (4.11) can be rewritten as

$$|\Delta f| \approx \frac{\Delta\varphi}{360 \cdot \tau}. \tag{4.19}$$

As mentioned before, the phase shift should be smaller than 180°, which means

$$|\Delta f| < \frac{180°}{360° \cdot \tau} = \frac{0.5}{\tau}. \tag{4.20}$$

Since $\Delta\varphi$ should not reach 180°, and it is impossible to predict its value before a measurement, this relation can only be treated as a rule of thumb for an upper bound. Another parameter regarding the measurement accuracy is phase uncertainty $\Delta\varphi_{un}$. It is mainly caused by the quantization errors during an analog-to-digital conversion. [VAN97] gives a closed expression for calculating this phase error. However, only a sinusoidal signal with a single frequency has been treated in this analysis. The results cannot be applied to the phase difference between two frequencies directly. In order to get a quick view, simulations with all possible apertures and phase shifts have been made, and the results show that the worst case uncertainty of the phase shift is about 0.05° and is independent of the aperture. The expected group delay uncertainty is given by

$$\tau_{un} \approx \frac{1}{360°} \cdot \left|\frac{\Delta\varphi_{un}}{\Delta f}\right| \leq \frac{1}{7200} \cdot \frac{1}{|\Delta f|}. \tag{4.21}$$

Regarding the frequency resolution of interest, it is reasonable to consider the carrier frequencies of the aforementioned modulation techniques. G3 uses 36 subcarriers, and the frequency distance is 1.5625 kHz. Within PRIME there is a spacing of 488.28125 Hz between adjacent carriers. It is desirable to set the aperture comparable to the smallest value. Finally, an aperture of 250 Hz provides an appropriate resolution. The quantization error is smaller than 0.56 µs. According to (4.20), this frequency resolution can be suitable for measuring a group delay up to 2 ms.

4.2.6.2 System Error

It can be seen in the measurement setup that the transmitted signal is recorded at both the transmitter and the receiver. Therefore, any difference between the two signal-pickup paths can lead to a measurement error. Fortunately, this kind of error is time-invariant. Thus, it can be determined in advance and then compensated within signal processing procedures.

4.2.6.3 Jitter Error of Synchronization

Recall that the GPS module has a synchronization uncertainty of 0.1 µs. The uncertainty due to synchronization jitter is

$$\frac{\tau_{syn}}{\mu s} \approx \frac{1}{360°} \cdot \left| \frac{360° \cdot \frac{\Delta f}{kHz} \cdot 0.1}{\frac{\Delta f}{kHz}} \right| = 0.1. \tag{4.22}$$

4.2.6.4 Error Caused by Channel Noise

The influence of noise on the amplitude and the phase can be expressed by

$$\sigma_H^{dB} = \frac{20}{\ln(10)} \cdot \sqrt{\frac{\sigma_{Tx}^2}{A_{Tx}^2} + \frac{\sigma_{Rx}^2}{A_{Tx}^2}}, \tag{4.23}$$

$$\sigma_\varphi^d = \frac{180}{\pi} \cdot \sqrt{\frac{\sigma_{Tx}^2}{A_{Tx}^2} + \frac{\sigma_{Rx}^2}{A_{Tx}^2}} \tag{4.24}$$

respectively. The standard deviation of the group delay is

$$\sigma_\tau = \frac{\sqrt{2}\sigma_\varphi}{\Delta f}, \tag{4.25}$$

where σ_H^{dB} and σ_φ^{d} denote the standard deviations of the magnitude in dB and the phase in degrees respectively. A_{Tx} and A_{Rx} are the spectral magnitudes at a given frequency for the transmitted and received signals respectively. σ_{Tx} and σ_{Rx} are standard deviations of noise at the given frequency for the transmitter and the receiver respectively. The noise is based on a zero-mean, uncorrelated noise model [ROL89]. The transmitter side is generally dominated by sounding signals, thus the SNR is relatively high. In channels with high attenuation the signal amplitude at the receiver side is small, so that the SNR can become very low. In this case, the deviations can be approximated by

$$\sigma_H^{dB} \cong \frac{20}{\ln 10} \cdot \frac{\sigma_{Rx}}{A_{Tx}} \tag{4.26}$$

and

$$\sigma_\varphi^{d} \cong \frac{180}{\pi} \cdot \frac{\sigma_{Rx}}{A_{Tx}}. \tag{4.27}$$

Therefore, the standard deviation of the group delay is

$$\sigma_\tau \cong \frac{\sqrt{2}}{\Delta f} \cdot \frac{\sigma_{Rx}}{A_{Rx}}. \tag{4.28}$$

4.2.7 Measurement Results

This part shows a measurement performed for a point-to-point channel in a university building. The measurement setups are connected to the same line phase. Fig. 4.3 shows the instantaneous CTFs for both directions. They are obtained by considering the first segments directly behind the zero-crossings of the mains voltage. The magnitude responses shown in plot (a) and the group delays shown in (c) are directly measured results, and the phase responses in plot (b) are reconstructed according to (4.13). The fact that the CTF for one direction is different from the CTF for the other direction indicates a un-symmetry of the observed channel. However, the difference is small and both directions have very similar traces. This is quite different from results obtained in the access domain.

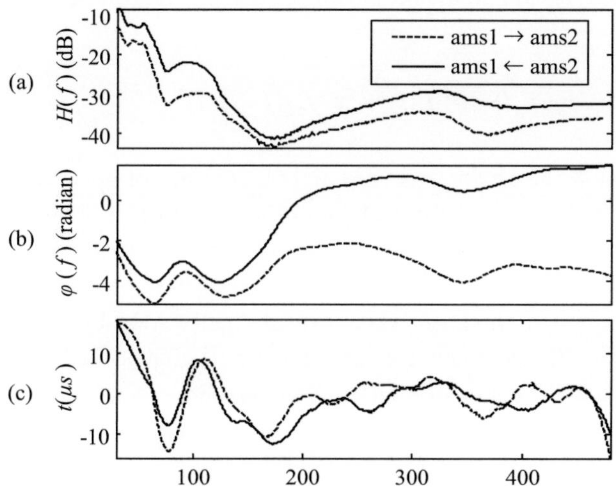

Fig. 4.3 Instantaneous CTFs for a bidirectional indoor channel. (a) amplitude responses. (b) phase responses. (c) group delays.

[SIG11] has reported measurements for three bidirectional channels in the access domain. Two of the channels exhibit obvious un-symmetry in the magnitude responses. It is impossible to find any similarity in the traces. The different degrees of un-symmetry could be explained by the physical lengths of these channels. For the indoor channel, the cable lengths are usually in the range between 10 to 50 meters. The cable cannot decouple the electric appliances connected at the measurement locations. The channel characteristics are determined by the total number of connected appliances, being visible from both directions. On the other hand, the decoupling effect of the cable plays a role in the access domain. The frequency selectivity of the CTFs is mainly "shaped" by the cable topologies and the appliances that are close to the individual measurement devices. Since different locations can have quite different local appliance scenarios, the two directions of a link can have different CTFs, despite the same network topology they are sharing. In addition to the un-symmetry, the group delay is not constant for both directions, thus the channel exhibits nonlinear phase responses. Equation (4.16) enables an investigation of the time-varying behavior of the phase at individual carrier frequencies. However, $\theta(f_c,l)$ is a superposition of phase details and a linear component. The linear component is caused by the sliding of the window function in the time domain. It is necessary to subtract the linear component from $\theta(f_c,l)$, so that the detailed phase fluctuation can be observed, according to

$$\theta_d(f_c,l) = \theta(f_c,l) - 2\pi \cdot f_c \cdot (l-1) \cdot T_o . \tag{4.29}$$

f_c and T_o denote the carrier frequency and the step size of the window shifting respectively. In Fig. 4.4 (a) the evolution of θ_d over time is shown at two carrier frequencies f_{c_1} and f_{c_2} respectively. Obviously, both traces show similar cyclic patterns. Taking the trace at f_{c_1} for example, the phase shows periodic changes. Within each period, the phase value experiences two local maxima. The difference between the maximum and the minimum is about $10°$. Fig. 4.4 (b) shows the evolution of the phase difference $\Delta\theta_d$ defined by

$$\Delta\theta_d(l) = \theta(f_{c_1},l) - \theta(f_{c_2},l). \tag{4.30}$$

Obviously, it changes with the same rate as the two phases. By comparing the phase evolutions at different frequencies, it has been observed that the fluctua-

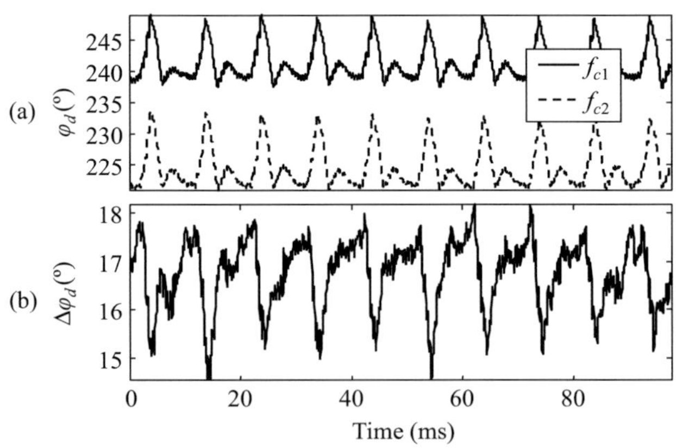

Fig. 4.4 Evolution of phase details at two carrier frequencies f_{c_1} and f_{c_2}.

tions of the phase and the group delay occur mostly below 300 kHz. In this range, the variations are usually larger at lower frequencies than at higher frequencies. At some frequencies the phase variations can exceed 50°. Similar large phase shift has also been reported at 109.375 kHz in [SUG09].

4.3 Modeling Complex-Valued CTFs

In [BAU05] measured amplitude responses using polynomial functions with a maximum order of 17 have been estimated. This method has only three parameters. However, there is weak correspondence with reality. Furthermore, the phase response has not been considered in the model. In [PHI99] a local maximum or minimum - also called a negative extreme - has been modeled with a series resonant circuit (SRC) which is composed of a resistor, a capacitor and an inductor. The channel can then be modeled by a cascade of such decoupled SRCs. This approach shows acceptable accordance with reality. Many electrical appliances have bypass capacitors at their power supply inputs. The feeding lines between these capacitors and the access points show resistive and inductive portions which can be considered as a serial combination of a resistor and an inductor. With the SRC model it is possible to map both the amplitude and the phase responses. The author has shown a relatively high accuracy of the SRC model for the high frequency range. Nevertheless, the SRC model has difficulties to describe the positive extremes in NB-PLC channels. In this thesis both the SRC and a parallel resonant circuit (PRC) are applied to model the extreme values. The complex-valued CTF is obtained by cascading all resonant circuits, as shown in Fig. 4.5, and then calculating the corresponding transfer function.

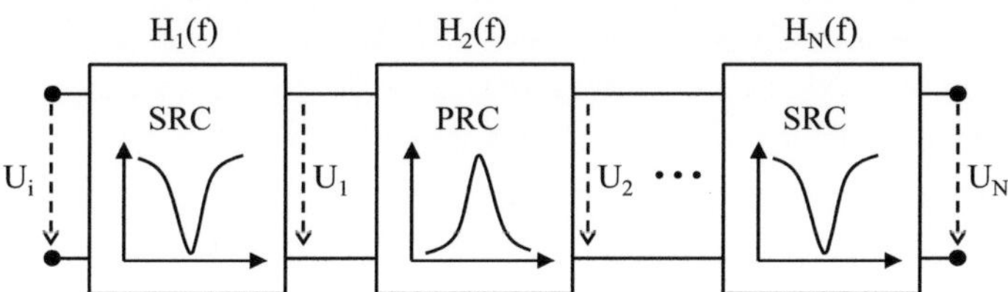

Fig. 4.5 Modeling a CTF using cascaded SRCs and PRCs.

4.3.1 Non-Loaded Resonant Circuit Models

Fig. 4.6 shows the SRC and the PRC in part (a) and (b) respectively. In the SRC circuit, a resistor R_S, a capacitor C_S and an inductor L_S are connected in series. L_S and C_S form an ideal series resonance circuit with a sharp notch. R_S damps the resonance and widens the notch.

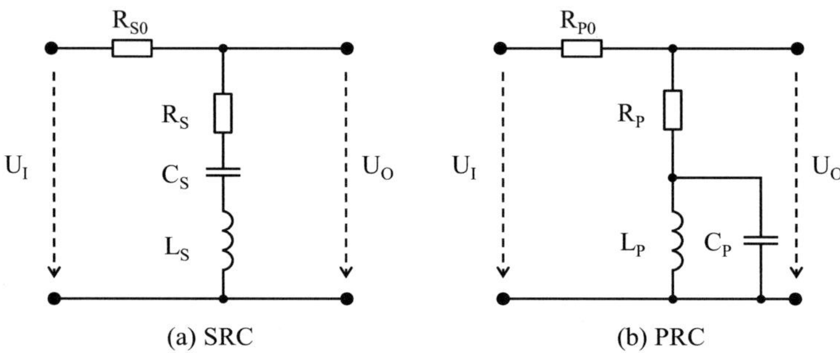

Fig. 4.6 Fundamental structure of (a) SRC and (b) PRC.

In addition, a resistor R_{S0} precedes the SRC. It models the line impedance of the powerline network which attenuates the transmitted signal and decouples the SRC from previous circuit stages. The PRC has the inductor in parallel with the capacitor. This PRC model applies for frequencies much higher than the mains frequency. An additional capacitor can be attached in series with L_P to extend this model to lower frequencies. The parallel resonance also has its origins in electrical appliances and devices, in particular in the transformers contained in many AC power supply units. The leakage inductance of the coils and the stray capacitance across the large mutual coil inductance dominate at high frequencies. They are connected in series, with one or more blocking capacitors in parallel. This circuit structure can exhibit a parallel resonance in the frequency range of interest. The impedance of the damped SRC can be described by

$$Z_S(f) = R_S + j \cdot 2\pi \cdot f \cdot L_S - j \cdot \frac{1}{2\pi \cdot f \cdot C_S}, \tag{4.31}$$

and the impedance of the PRC is

$$Z_P(f) = R_P + \frac{\dfrac{L_P}{C_P}}{j \cdot 2\pi \cdot f \cdot L_P - j \cdot \dfrac{1}{2\pi \cdot f \cdot C_P}}. \tag{4.32}$$

In both circuits, the resonant frequency is determined by the inductance L and capacitance C respectively, i.e.

$$f_{res} = \frac{1}{2\pi \cdot \sqrt{L_P C_P}} \quad or \quad f_{res} = \frac{1}{2\pi \cdot \sqrt{L_S C_S}}. \tag{4.33}$$

The transfer function of each circuit model is determined by the impedance Z_S or Z_P and the series resistor R_{S0} or R_{P0} respectively, thus it can be obtained as

$$H_S(f) = \frac{Z_S}{R_{S0} + Z_S} \quad or \quad H_P(f) = \frac{Z_P}{R_{P0} + Z_P}. \tag{4.34}$$

With given values of R_{S0} (R_{P0}) and f_{res}, the attenuation and the shape of the notches are controlled by R_S (R_P) and the ratio

$$\beta = \frac{C_S}{L_S} \quad or \quad \frac{C_P}{L_P}. \tag{4.35}$$

Fig. 4.7 and Fig. 4.8 correspond to the SRC. In Fig. 4.7, both the attenuation at f_{res} and the peak values of the phase on both sides of the resonant frequency decrease with increasing R_S. The phase change from the negative to the positive extremes is decelerated by a higher R_S. In Fig. 4.8, the peak values of the attenuation and the phase are not affected by different β values. The sharpness of the notch is reduced by an increasing β.

In Fig. 4.9 it can be seen that R_P doesn't affect the 3-dB attenuation significantly. Instead, R_P determines the attenuation values at frequencies far away from the resonance frequency. Similar to the situation in the SRC, β in Fig. 4.10 affects the shape of the extreme in the PRC. Nevertheless, the 3-dB bandwidth increases with a reduced β value.

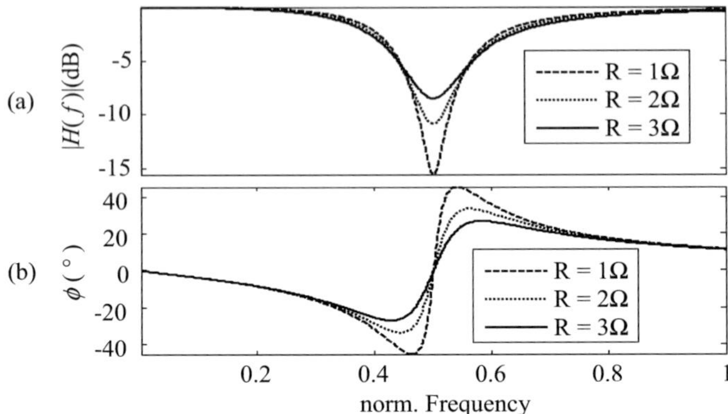

Fig. 4.7 CTFs of SRC for different values of R_S.

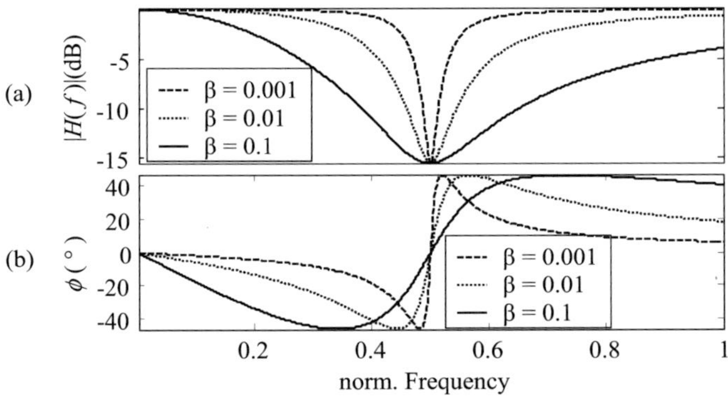

Fig. 4.8 CTFs of SRC with different β.

4.3.2 Calculation of Components for Non-Loaded Models

This section introduces a two-point approach to calculate all three components for a SRC or PRC model. Suppose only the measurement result of the magnitude response is available. Such an assumption has its background in practice because many channel measurements have been limited to the magnitude response.

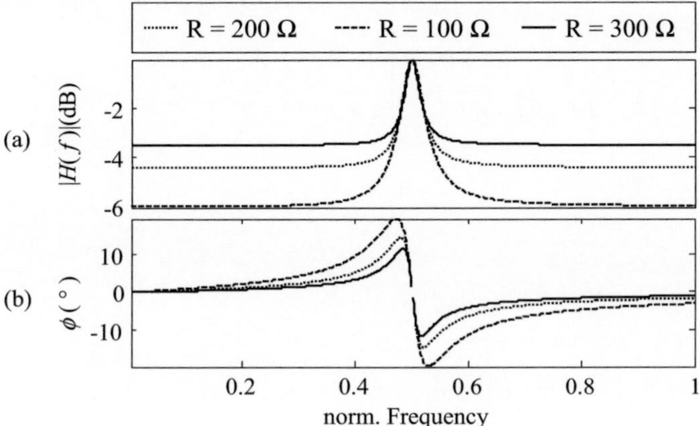

Fig. 4.9 CTFs of PRC with different values of R_P.

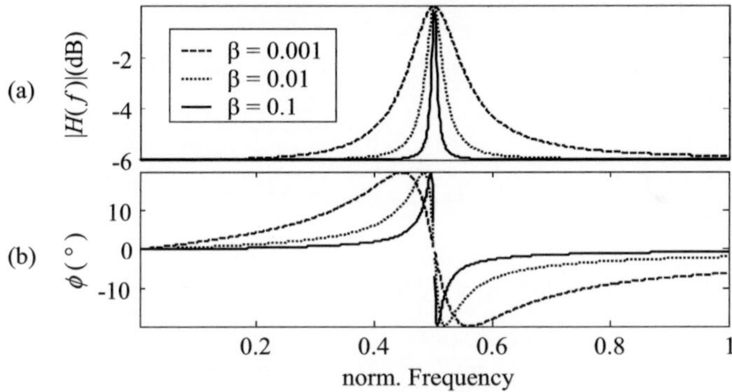

Fig. 4.10 CTFs of PRC with different values of β.

For the SRC, two points $(f_{res}, |H_{res}|)$, $(f_1, |H_1|)$ can be obtained from such measurement results directly:

$$R_S = R_{S0} \cdot \frac{|H_{res}|}{1 - |H_{res}|}. \tag{4.36}$$

Considering (4.31), (4.33) and (4.34) , the capacitance C_S can be obtained by

$$C_S = \frac{1}{2\pi \cdot f_1} \sqrt{\frac{\left(1 - |H_1|^2\right) \cdot \left(1 - f_1^2 / f_{res}^2\right)^2}{|H_1|^2 \cdot \left(R_{S0} + R_S\right)^2 - R_S^2}} \, . \tag{4.37}$$

The inductance L_S is

$$L_S = \frac{1}{\left(2\pi \cdot f_{res}\right)^2 \cdot C_S} \, . \tag{4.38}$$

As shown in Fig. 4.9, there is almost no attenuation at f_{res}. The flat attenuation at frequencies far away from f_{res} is mainly determined by R_{P0} and R_P, i.e.

$$\hat{H}_\infty = \frac{R_P}{R_{P0} + R_P} \, , \tag{4.39}$$

where $\hat{H}_\infty$ is defined as

$$\hat{H}_\infty = \frac{|H(f_\infty)|}{|H(f_{res})|} \, . \tag{4.40}$$

Suppose R_{P0} is a pre-defined value. Then R_P can be obtained by

$$R_P = R_P \cdot \frac{\hat{H}_\infty}{1 - \hat{H}_\infty} \, . \tag{4.41}$$

Using (4.32), (4.33) and (4.34), L_P can be obtained by

$$L_P = \frac{\left|1 - f_1^2 / f_{res}^2\right|}{2\pi \cdot f_1} \cdot \sqrt{\frac{\hat{H}_1^2 \cdot \left(R_{P0} + R_P\right)^2 - R_P^2}{1 - \hat{H}_1^2}} \, , \tag{4.42}$$

where f_1 is the frequency next to the resonance, and $\hat{H}_1$ is defined by

$$\hat{H}_1 = \frac{|H(f_1)|}{|H(f_{res})|}.$$ (4.43)

To obtain a valid value for L_P, it is important to keep the value inside a positive range of the square root in (4.42), which means

$$\hat{H}_1 < 1,$$ (4.44)

and

$$\hat{H}_1 \cdot (R_{P0} + R_P) > R_P.$$ (4.45)

The condition in (4.65) can be fulfilled by assuring f_1 to be different from f_{res}. Meanwhile, (4.45) can be rewritten as

$$\hat{H}_1 > \frac{R_P}{R_{P0} + R_P} = \hat{H}_\infty.$$ (4.46)

It can be seen in Fig. 4.9 and Fig. 4.10 that $|H(f)|$ decreases with an increasing distance between f and f_{res}. Therefore, f_1 shall not be too far away from f_{res}, so that (4.45) can be satisfied. The last parameter C_P is obtained by

$$C_P = \frac{1}{\left(2\pi \cdot f_{res}\right)^2 \cdot L_P}.$$ (4.47)

It can be seen that two measurement points are sufficient to estimate the components for the fundamental SRC and PRC models.

4.3.3 Influence of Loads on Circuit Models

The aforementioned PRC and SRC models have been treated as non-loaded circuits. However, in the cascaded structure shown in Fig. 4.11, the actual transfer function $H_{Lm}(f)$ of the mth circuit model is affected by the equivalent impedance $Z_{m+1}(f)$ formed by all circuits behind the model.

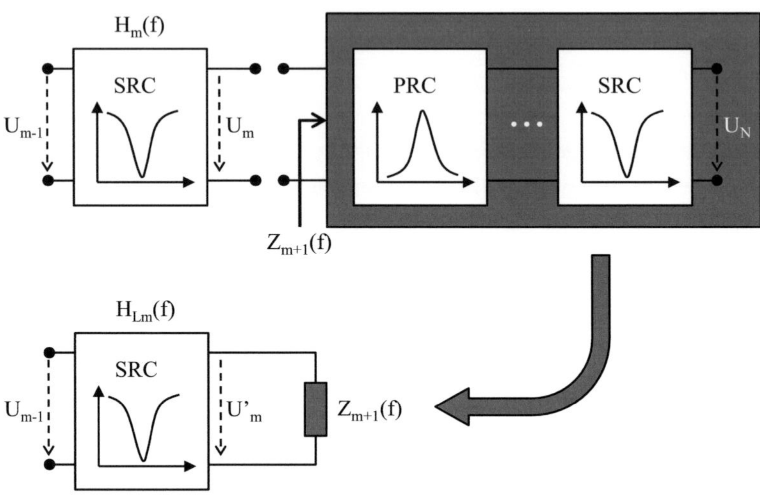

Fig. 4.11 The actual transfer function of mth model is affected by $Z_{m+1}(f)$.

In order to determine the influence of $Z_{m+1}(f)$ on $H_m(f)$ quantitatively, $Z_{m+1}(f)$ is added in the equivalent circuit of the non-loaded SRC model, as shown in Fig. 4.12.

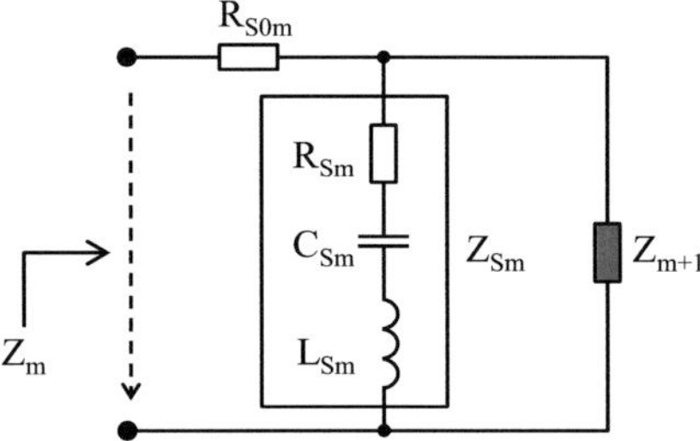

Fig. 4.12 Equivalent circuit of the loaded SRC model.

Z_{Sm} and Z_{m+1} are connected in parallel now. The total impedance of this parallel arrangement, denoted by Z_{SLm}, can be obtained by

$$Z_{SLm} = \frac{Z_{m+1} \cdot Z_{Sm}}{Z_{m+1} + Z_{Sm}}. \tag{4.48}$$

Equation (4.34) is modified to

$$H_{Lm}(f) = \frac{Z_{SLm}}{R_{S0m} + Z_{SLm}},$$

(4.49)

By replacing Z_{SLm} in (4.49) with (4.48), the relation between $H_{Lm}(f)$ and $H_m(f)$ can be expressed as

$$H_m(f) = \frac{1}{\dfrac{1}{H_{Lm}(f)} - \dfrac{R_{S0m}}{Z_{m+1}}}.$$

(4.50)

Obviously, it is possible to extract the non-loaded transfer function $H_m(f)$ out of a loaded $H_{Lm}(f)$ if R_{S0m} and Z_{m+1} are already known. The model parameters R_{Sm}, C_{Sm} and L_{Sm} can be estimated using the method introduced in 4.3.2.

After having obtained all three parameters, the equivalent input impedance looking in to the circuit can be obtained by

$$Z_m = R_{S0m} + Z_{SLm}$$

(4.51)

4.3.4 From Amplitude Response to Model Parameters

Based on the two-point approach, this section proposes an iterative algorithm to model the CTF by using cascaded SRCs and PRCs. The idea is to decompose the measured magnitude response $|H(f)|$ into multiple fundamental magnitude responses $|H_m(f)|$ iteratively. Each fundamental magnitude response contains only one extreme and refers to a single PRC or SRC circuit. Based on the cascaded structure shown in Fig. 4.5, the decomposition is carried out from right to left ($m=N$, $N-1$,..., 1), i.e. $H_N(f)$ is estimated first, followed by $H_{N-1}(f)$, $H_{N-2}(f)$ and so on until parameters for all N models are determined. Finally the modeled CTF is synthesized by cascading all these resonance circuits and calculating the overall transfer function. In this way, a complex-valued CTF can be obtained from the measured magnitude response $|H(f)|$.

In the first step of the mth iteration, the load Z_{m+1} for the mth circuit is calculated by applying Z_{m+2} and the parameters of the $(m+1)^{\text{th}}$ circuit model in (4.48) and

(4.51). For the first iteration where $m = N$, the load corresponds to the input impedance of a PLC receiver and has usually a high impedance value. In this case, a value of 100 Ω can be applied for simplicity. A constant value, for example 5 Ω, can be assigned to R_{S0m}, so that the number of unknown parameters is reduced. By applying Z_{m+1} and R_{S0m} in (4.50) the influence of Z_{m+1} can be got rid of the loaded circuit model. The resulted magnitude response, denoted by $|H(f)|$, can be used to estimate the parameters of the mth non-loaded model.

The extreme with the highest magnitude in $|H(f)|$ is located. The corresponding frequency is the resonant frequency f_{res}. Fig. 4.13 shows the magnitude response for a SRC. A negative extreme at around 170 kHz is considered to be caused by a SRC.

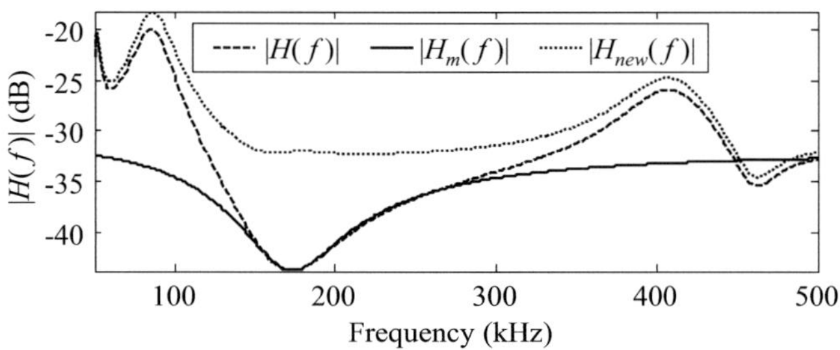

Fig. 4.13 Eliminating negative extremes using the SRC model.

Fig. 4.14 shows the magnitude responses for a PRC. It has a positive extreme at about 400 kHz. Four values: f_{res}, $|H(f_{res})|$, a second frequency f_1 in the vicinity, and the corresponding magnitude $|H(f_1)|$ are recorded. For a SRC model, the cir-

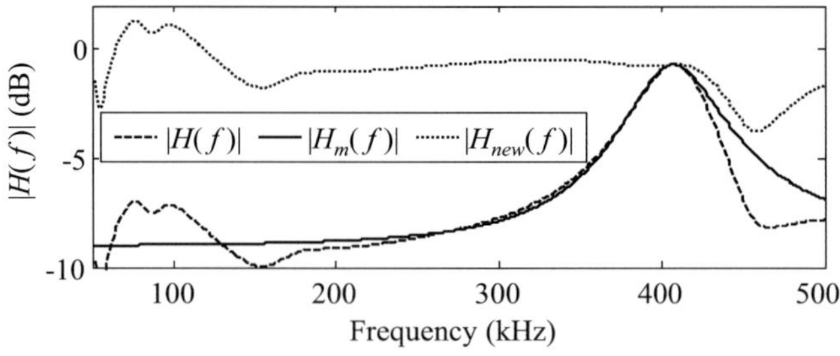

Fig. 4.14 Eliminating positive extremes using the PRC model.

cuit components R_S, L_S and C_S can be obtained by using (4.36), (4.37) and (4.38). For a PRC model, the magnitude values $|H(f_{res})|$, and $|H(f_1)|$ cannot be used directly. Instead, the relative magnitudes $\hat{H}_\infty$ and $\hat{H}_1$ shall be calculated according to (4.40) and (4.43) respectively. Then the circuit components R_P, L_P and C_P can be obtained using (4.41), (4.42) and (4.47).

In the next step, the complex transfer function $H_m(f)$, is synthesized for the current fundamental circuit using (4.34). It can be seen that the peak of $|H_m(f)|$ matches the targeted extreme in $|H(f)|$ very well. $|H(f)|$ is than divided by $|H_m(f)|$, so that the current extreme can be eliminated. The resulted new magnitude response, denoted $|H_{new}(f)|$ is used as the residual loaded transfer function in the next iteration. Till now, the mth circuit model is identified, and the mth iteration can be terminated. With respect to the selection of f_1, different values result in different shapes of $|H_m(f)|$, and thus, different values of $|H_{new}(f)|$. It could be necessary to try multiple values of f_1 to achieve an appropriate result. The whole process can be terminated as soon as no extreme peak exceeds a predefined threshold, for example 3 dB. Fig. 4.15 gives an example of measured and synthesized magnitude responses in (a). Plot (b) refers to the measured and synthesized group delays. The synthesis matches the measurement results quite well, despite deviations at some frequencies. The accuracy can be improved by reducing the threshold. This has a direct consequence in that the number of fundamental reso-

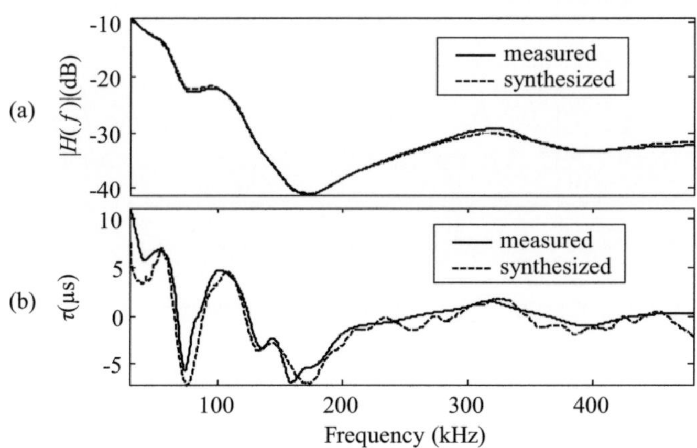

Fig. 4.15 Measured and synthesized transfer functions. (a) measured and synthesized magnitude responses (b) measured and synthesized group delays.

nance circuits will increase. Generally speaking, the number of SRCs and PRCs determines the accuracy and the complexity. Thus, a compromise is always recommended in each application.

4.4 Emulation of the Transfer Function

This section deals with the emulation of CTFs. As mentioned earlier, the mathematical interpretation of the influence on the transmitted signal is a convolution of this signal with the channel's impulse response. Thus, to emulate a CTF means to implement a convolutional algorithm. A starting point is a known CTF including magnitude, phase response, and group delay, denoted by $|H_{dB}(f)|$, $\varphi(f)$ and $\tau(f)$ respectively. Fig. 4.16 shows an example of a CTF which will be used as the target CTF in the following. This CTF is not from a real-world measurement, but was created artificially. Sharp turns and edges are intentionally inserted to examine the performance of the emulation algorithm thoroughly.

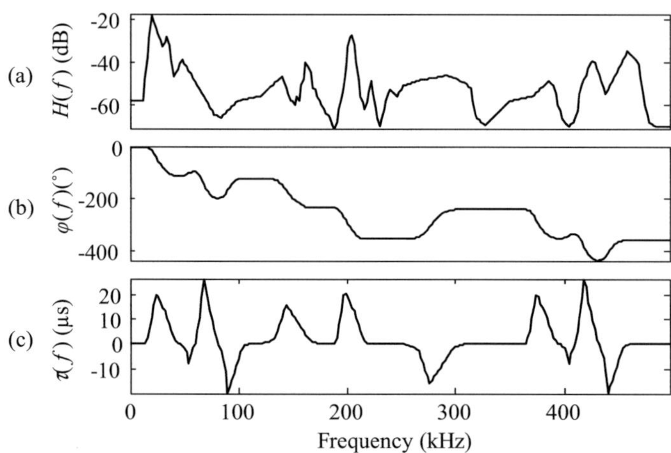

Fig. 4.16 Magnitude, phase response and group delay of a desired CTF.

4.4.1 Direct Convolution and FFT-based Convolution

There are two possibilities to implement convolutional algorithms on FPGAs. The first one is a conventional time-domain approach using a digital filter, such as a finite impulse response (FIR). The other is a block convolution in the frequency domain.

The first method obtains the channel impulse response and saves them as coefficients of the FIR filter. The transmitted signal is correlated with the filter coefficients. The block convolution is based on the principle that a convolution in the time domain corresponds to a multiplication in the frequency domain. It transforms the transmitted signal into the frequency domain by a DFT, multiplies the spectrum with the CTF, and transforms the product back into the time domain via IDFT. This method takes advantages of highly efficient fast Fourier transform (FFT) algorithms, and is also called 'FFT convolution'. By using an FFT algorithm to calculate the DFT, this kind of convolution can be faster than the direct convolution in the time domain. Common methods to implement the FFT convolution are the overlap-add, the weighted overlap-add and the overlap-save methods. Details of these methods can be found in [RAB75]. By comparing the number of major operations it has been realized that the overlap-add method can be faster and more efficient than the FIR-based time-domain convolution if the filter order exceeds 20 or 40 for complex- and real-valued filters respectively. One of the disadvantages of the FFT convolution is an inherent delay within the signal processing chain. To investigate this delay in detail we use an overlap-add-based system reported in [CAN08] as an example. The system is using a N_{FFT} point FFT convolution. Fig. 4.17 illustrates a simplified scheme. The system clock cycles which the FPGA needs in each module and the corresponding redundancies are denoted by N_1 through N_4 and τ_1 through τ_4 respectively. The channel is supposed to have an impulse response of N_{ir} samples. It is padded with $(N_{FFT} - N_{ir})$ zeros, converted to the frequency response H. The input signal is converted to digital samples by the ADC. The "overlap-add input" module collects $(N_{FFT} - N_{ir})$ samples, adds N_{ir} zeros to build a FFT block and delivers this block to the FFT

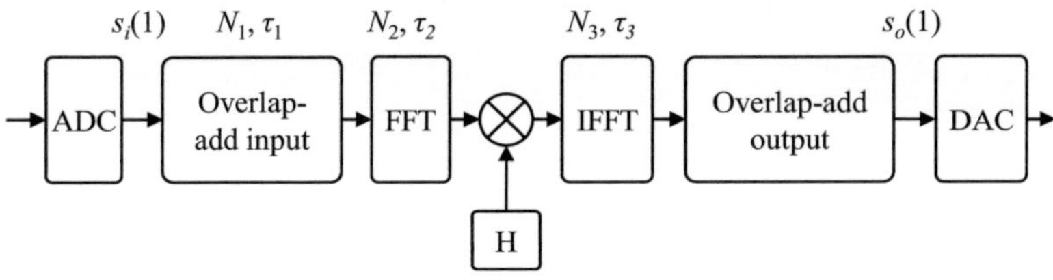

Fig. 4.17 Implementation of the overlap-add method.

module. The first sample value $s_i(1)$ must wait for at least N_1 sampling periods until it can be processed by the next module, whereby

$$N_1 = N_{FFT} - N_{ir},\tag{4.52}$$

The FFT module performs a N_{FFT}-point FFT on the block. The FPGA needs N_2 clock cycles to finish the FFT. This clock frequency can be different from the sampling rate. The converted values are transferred to the next module one by one. At the same time, they are multiplied by the corresponding values of H. In this case, the multiplication needs at least N_{FFT} clock cycles. The product vector is converted to a waveform in the time domain by an IFFT which needs the same number of clock cycles as the FFT module. The "overlap-add output" module adds the current IFFT block to the previous one with an appropriate superimposition. Finally, the output value corresponding to $s_i(1)$ is obtained.

The delays caused by each module are listed in Table 4.1. According to the table, the proposed system uses a 4096 point FFT and IFFT respectively. The signal is sampled at 100 MSps. The channel impulse response is supposed to be 15.36 µs. Each signal block contains 2560 samples, leading to a latency of 2560 sample periods in the input module. The FPGA needs 12,445 clock cycles for calculating the FFT or IFFT [XIL11]. It is assumed that the FFT, the IFFT and the multiplications are performed at the maximum clock rate, for example at 395 MHz. As a result, the latencies of FFT and IFFT are both 31.51 µs, and the multiplication takes 10.37 µs. The minimum total latency reaches 99 µs, being more than 6 times longer than the channel impulse response. Applying the same conditions for convolution with a sample rate of 2 MSps, the total latency would be about 1.35 ms. Note that this large latency is not caused by the physical channel, but by the convolution hardware. It may, however, not impair the data transmis-

Table 4.1 Delays Caused by the Modules in Fig. 4.17

N_{FFT}	N_{ir}	N_1	τ_1 (µs)	N_2	τ_2 (µs)	N_3	τ_3 (µs)	N_4	τ_4 (µs)
4096	1536	2560	25.6 (100M) / 1280 (2M)	12445	31.51 @ clock rate of 395MHz	4096	10.37 @ clock rate of 395MHz	12445	31.51 @ clock rate of 395MHz

sion if the system deploys a preamble for frame synchronization, such as the OFDM or DCSK systems mentioned in chapter 2, because both the preamble and the transmitted signal are delayed by the same amount. However, there are systems that use the zero-crossings of the mains voltage for bit and symbol synchronization, for example, most S-FSK and other single-carrier systems [IEC613], as well as some OFDM systems [KIS08]. The latency in the convolution can introduce an "artificial" synchronization error which degrades the data transmission quality in an unwanted manner. Therefore, the procedure proposed above is not suitable for a universal testbed for NB-PLC systems.

In contrast to the FFT-convolution, the direct convolution using a FIR filter does not have an inherent delay. After evaluating both approaches, the direct convolution has been selected as appropriate method for this thesis.

4.4.2 FIR Filter Method

Nowadays, digital filters are widely used. The rapid development of modern high-speed integrated circuits on silicon makes the implementation of fast and complex filter structures possible, even for challenging real-time applications. The design of optimized filter structures and the improvement of the corresponding operation speed have been well studied and developed in the recent years. Many manufacturers provide optimized intellectual property (IP) cores, so that the effort for hardware implementation can be widely reduced. Therefore, users can focus on their specific application of such digital filters. In this context, it appears not necessary for this thesis to go into details of filter design and implementation. Instead, major effort is spent to design proper filter kernels in order to emulate various CTFs with high accuracy.

4.4.3 Frequency Sampling Method

Among the approaches to design and optimize filter kernels, the windowed frequency sampling method is considered as the simplest and most straightforward one. The filter coefficients can be obtained by performing an IFFT on the desired frequency response. This procedure is quite attractive since it can be applied to arbitrary courses of CTFs. Suppose the filter has a length of N_{fir}, then the filter has an order of $N_{fir}-1$. The frequency response is then expressed in linear scale by

$$H(i) = 10^{\frac{\left|H_{dB}(i)\right|}{20}} \cdot \exp\left[j\varphi(i)\right].$$ (4.53)

The input vector for the IFFT should be filled carefully, so that the IFFT result will be real-valued. For this purpose, the real and the imaginary parts of $H(i)$ shall have an even and odd symmetry respectively [SMI99]. These symmetries are illustrated in Fig. 4.18, where the zero-frequency bin is shifted to the center. The real and imaginary parts are both normalized and the plot has been zoomed in, so that the significant values of both parts can be seen clearly.

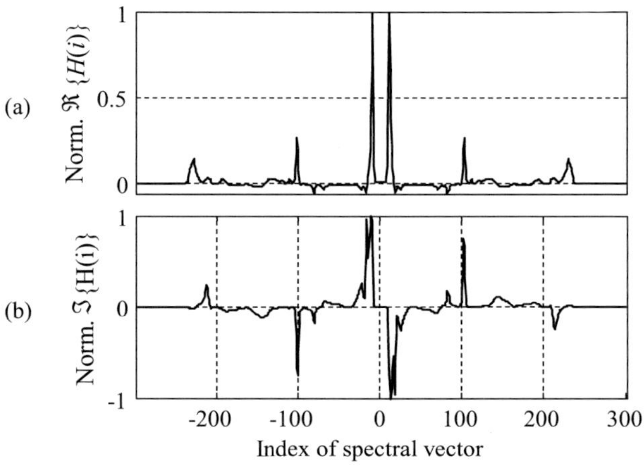

Fig. 4.18 Frequency domain vector: (a) even symmetry for real part. (b) odd symmetry for imaginary part. The vector has 1000 samples, and the plot is enlarged so that the frequency interval with significant energy, i.e. between -240 and 240, is clearly visible.

The result of the IFFT shown in Fig. 4.19 cannot be used as a filter kernel directly. It may cause aliasing in the time domain during a convolution. Therefore, swapping and shifting are necessary to obtain a valid filter kernel [SMI99]. The swapping can be done by simply exchanging the left and right halves. Obviously, the swapped IFFT has significant side lobes. It does not move towards zero on both sides, which will lead to aliasing in the time domain and a loss of dynamic range of the filter response. To reduce these unwanted effects, the swapped IFFT vector can be multiplied by a window function. A window function has a bell-like shape with its maximum in the center, while both ends rapidly decline to zero. There are many window functions available, for example Tukey, Cosine, Blackman, Kaiser, Gaussian, and more. Here, the Kaiser window function is preferably used.

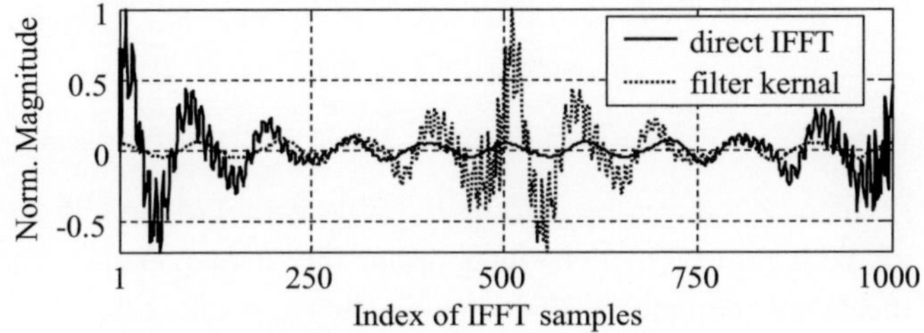

Fig. 4.19 Direct output of the IFFT and corresponding filter kernel.

The Kaiser window approximates prolate-spheroidal wave functions which try to maximize the energy in the frequency band of interest within restricted time duration [HAR78]. The Kaiser window is defined by

$$w(i) = \begin{cases} \dfrac{I_0\left[\pi\alpha\sqrt{1-\left(\dfrac{i}{N_{fir}/2}\right)^2}\,\right]}{I_0(\pi\alpha)}, & 0 \le |i| \le \dfrac{N_{fir}}{2}. \\ 0, & otherwise \end{cases} \qquad (4.54)$$

$I_0(x)$ is the zero-order modified Bessel function of the first kind, given by

$$I_0(x) = \sum_{k=0}^{\infty}\left[\frac{(x/2)^k}{k!}\right]^2. \qquad (4.55)$$

α is the parameter in (4.54) which defines the width of the window. In the frequency domain, it determines the attenuation of the side lobes. Fig. 4.20 shows the Kaiser window with α=32 and the windowed filter kernel. The side lobes are strongly reduced, while the most significant part between index 250 and 750 is well preserved. The original filter kernel itself causes a delay of $N_{fir}/2$. This delay,

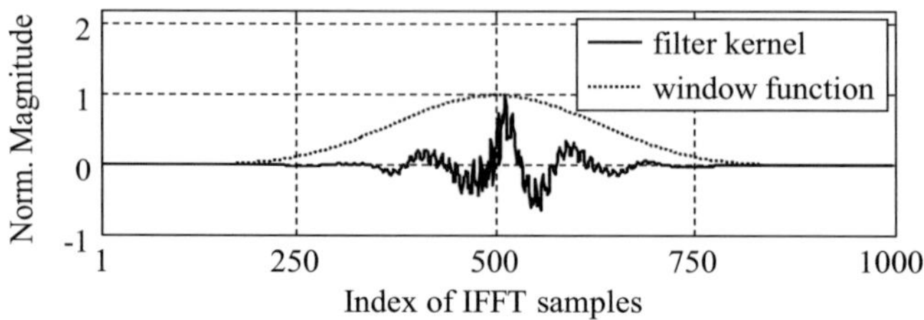

Fig. 4.20 Kaiser window function with $\alpha=32$ and windowed filter kernel.

denoted by t_L, can be reduced by shifting the filter kernel to the left and padding the right side with zeros. The delay is constant for all frequencies. The delayed version of the desired complex CTF is obtained by

$$H_{shift}(i) = H(i) \cdot \exp\left[j2\pi \cdot i \cdot \left(1 - \frac{N_{shift}}{N_{fir}} \right) \right], \tag{4.56}$$

where f_s denotes the sampling rate, and N_{shift} is the number of samples by which the CTF vector must be shifted.

$$N_{shift} = \left\lfloor \frac{t_L}{f_s} \right\rfloor. \tag{4.57}$$

The filter kernel should also be shifted to the left by N_{shift} samples in the time domain, so that the resulting filter kernel is

$$h_{shift}(i) = \begin{cases} h(i + N_{shift}), & 1 \leq i \leq N_{fir} - N_{shift} \\ 0, & N_{fir} - N_{shift} < i \leq N_{fir} \end{cases}. \tag{4.58}$$

The operation in (4.58) discards the parts $h(1)$ through $h(N_{shift})$ and leads to a loss of signal energy. If the shifting is driven too much to the left or to the right, a large number of filter coefficients with significant energy may get lost. Then the deviation from the desired CTF becomes unacceptable - see Fig. 4.21. Here, the phase error reaches 35° at 199 kHz. The maximum relative error of the group delay exceeds 18%. Another problem is that the frequency response between two adjacent frequency samples is not constrained. Therefore, the resulting curve can deviate from the desired response between the samples. The obtained set of coefficients is then probably suboptimal [SMI11].

4.4.4 Extension of the Frequency Sampling Method

Sharp edges of the magnitude response can be smoothed by the aforementioned processing steps. After that, most parts of the magnitude response can be preserved. The overall situation seems to be less critical for real NB-PLC channels, because sharp edges, as shown in the first plot of Fig. 4.21, are quite rare in the frequency range of interest. Therefore, we can focus on reducing the errors in the phase and the group delay responses.

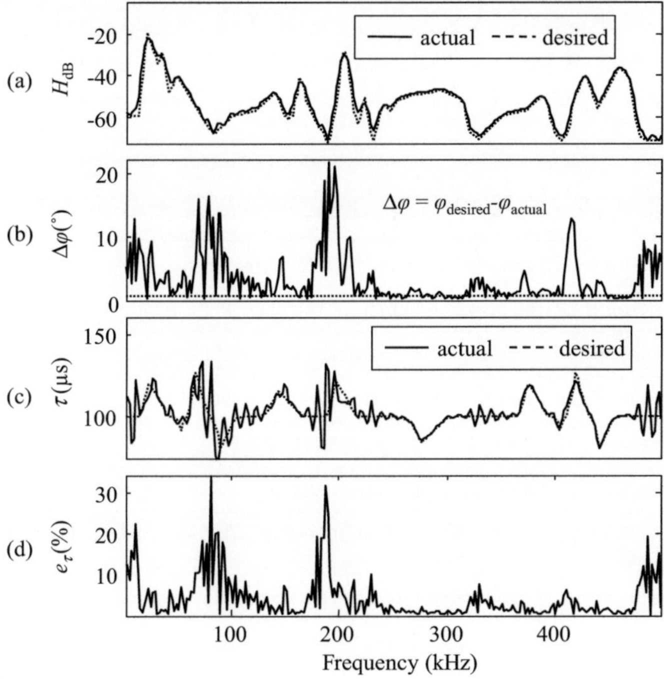

Fig. 4.21 Deviation from a desired CTF due to shifting, truncating and zero-padding.

An iterative algorithm is presented to reduce phase errors to a defined bound $e_b(f)$

$$\left| \Delta\varphi(f) \right| \le e_b(f). \tag{4.59}$$

The bound can be defined arbitrarily. In this thesis, it is defined by

$$e_b(f) = \frac{e_1}{|H(f)|^{e_2}} + e_3, \qquad (4.60)$$

where e_1, e_2 and e_3 are tunable parameters, $H(f)$ is the desired magnitude response in a linear scale. Obviously the bound is frequency dependent. The phase errors at frequencies with high attenuation are less critical than those at the other frequencies. Therefore, they are weighted with smaller factors. The parameter e_3 is a constant for all sampled frequencies and determines the offset component for the phase error bound. e_2 adjusts the bound difference between frequencies with high and low attenuation. A larger e_2 leads to a larger bound difference. e_1 determines the ratio of frequency-selective bound component to the overall bound component. The following steps are to be considered for the error reduction:

Table 4.2 Operation of Extended Frequency Sampling

Step	Operation
1	Perform shift and truncation on the filter kernel using (4.58), obtain h_{shift1}
2	Perform FFT on h_{shift1}, obtain H_{shift1}, calculate phase error $\Delta\varphi_1$
3	Multiply H_{shift1} by $\exp(j{\cdot}\Delta\varphi_1)$, perform IFFT, obtain h_{shift2}
4	Repeat step 1 and 2 once, calculate H_{shift2}, $\Delta\varphi_2$ and $|H/H_{shift2}|$
5	Multiply H_{shift2} by $|H/H_{shift2}|$, perform IFFT, obtain h_{shift3}
6	Repeat step 1 and 2, calculate H_{shift3} and $\Delta\varphi_3$
7	Check inequality (4.59), and find frequency bins f_x at which the inequality is not fulfilled
8	Multiply $H_{shift3}(f_x)$ by $\exp[j{\cdot}\Delta\varphi_3(f_x)]$, perform IFFT, obtain h_{shift4}
9	Repeat step 6, 7 and 8 until the phase errors at all frequencies fulfill (4.59)

Fig. 4.22 shows an example of the result. Compared with the plots shown in Fig. 4.21, the phase and group delay errors are largely reduced. Large errors of phase and group delay occur at frequencies around 16, 84, 188, 230, 328, 404 and 490 kHz, where the magnitude suffers from high attenuation and also features

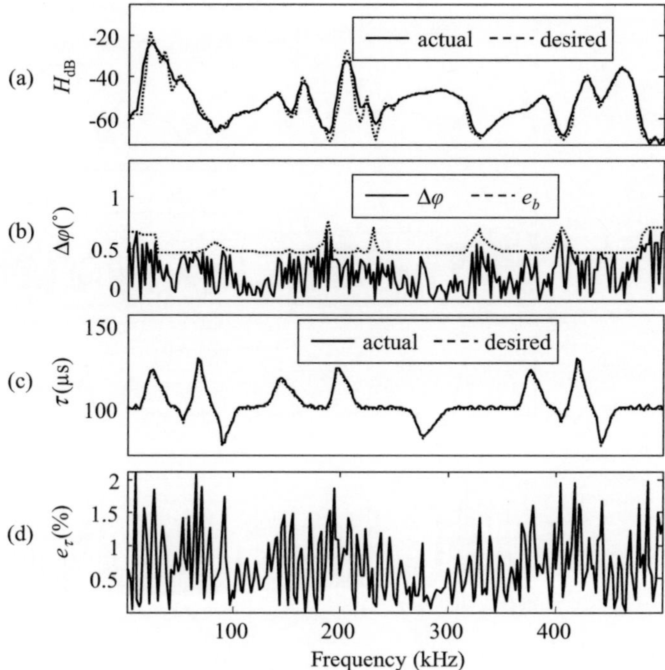

Fig. 4.22 Improved results by applying iterative error reduction, $e_1=10^{-9}$, $e_2=1.9$, $e_3=7.9\cdot10^{-3}$, $\mathrm{E}\{e(\tau)\}=1.71\%$.

sharp edges. The group delays corresponding to moderate and low magnitude attenuation are even lower than 1%. The expected value of the relative error is as low as 1.71%. This coefficient set has a relatively large e_2 in comparison with e_3.

The bound can also have a dominant e_3. As shown in Fig. 4.23, the bound is almost a constant over the whole frequency range. In this case, the peak values of the phase error corresponding to high attenuation are no longer clear to see. They seem to be distributed over a larger range of frequencies.

The iterative extension to the frequency sampling method provides a possibility to improve the emulation accuracy of arbitrary CTFs. In addition, the redundancy of the implemented filter can also be reduced to fulfill real-time requirements. It is suitable for both short and long FIR filters. Nevertheless, the obtained filter coefficients are not necessarily optimal. The user has to tune the parameters e_1

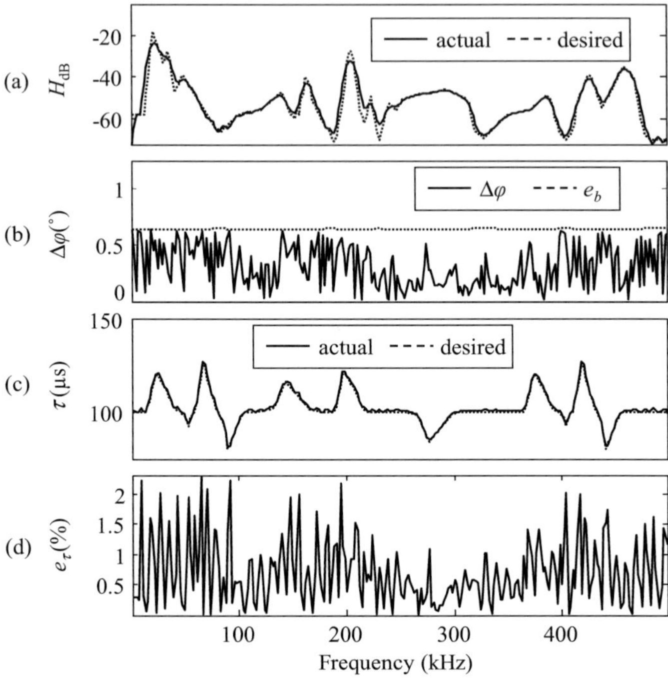

Fig. 4.23 Improved results by applying iterative error reduction, $e_1=10^{-6}$, $e_2=0.5$, $e_3=1.12\cdot10^{-2}$, $E\{e(\tau)\}=1.74\%$.

through e_3 to obtain satisfying coefficients. The quality also depends on the specification of the error bound. The iteration could run into convergence problems if the bound is not defined appropriately.

4.4.5 Hardware Implementation

The FIR filter is implemented in a Cyclone III FPGA. A fully integrated FIR filter development environment, a FIR compiler is also provided. The compiler supports several hardware architectures, such as the fully parallel filter, the fully serial filter, the multi-bit serial filter, and the multi-cycle variable (MSV) filter. The different versions fulfill different requirements on performance, resource utilization and speed. The first three structures are based on distributed arithmetic (DA) which utilizes memory units instead of multipliers to perform convolutions, correlations, DFT and so on [MEY07]. The fully parallel structure is composed of large amounts of memory units running in parallel. Thus, high speed can be achieved at the cost of silicon area. The fully serial structure minimizes the area by using only one memory unit. It calculates only a part of the result in each clock cycle. A complete computation requires multiple clock cycles. The multi-bit serial filter implements a moderate number of memory units. This is a trade-off between throughput and resource usage. In this variant, the throughput is higher than for the fully serial structure but lower than for the parallel implementation. The multi-cycle variable structure takes advantage of the embedded DSP block multipliers. The number of DSP blocks and the number of calculation clock cycles can be configured. The data throughput increases with reduced number of calculation cycles, while the area decreases as the number of cycles increases. Modern FPGAs have more and more embedded high-speed DSP blocks. In comparison with the DA-based structures, the MCV can achieve a similar maximum speed with much fewer memory units [ALT09]. Therefore, the MCV is preferred in the following parts. As mentioned before, the sampling rate is 2 MSps. To achieve a frequency resolution of 2 kHz, 1000 filter coefficients are needed. 50 cycles are needed for one filter output and the filter is driven by a 100 MHz clock. The coefficient reloading feature allows the change of filter coefficients during the convolution operation, and thus makes the implementation of time-variant filters possible.

4.4.6 Determination of Effective Bits[1]

The convolution of the transmitted signal with the filter coefficients is based on a multiply and accumulate (MAC) operation. The word length of the filter output can be calculated by

$$B_o = B_i + B_c - 1 + ceil\left(\log_2 N_c\right), \tag{4.61}$$

where B_o denotes the maximal total word length of the filter output, B_i and B_c are the width of the filter input and the filter coefficients respectively. Both of them are signed 12-bit values, thus resulting in a filter output word length of 33 bits. A

---

[1] This part has been published in [LIU11]: W. Liu, M. Sigle and K. Dostert, "Advanced Emulation of Channel Transfer Functions for Performance Evaluation of Powerline Modems," in Proc. IEEE Int. Symp. Power Line Commun. Applicat., Udine, Italy, Apr. 2011, pp. 446-451.

difficulty arises when the filtered values are converted to an analog signal by a DAC which is only 14-bits wide. As a result, 14 data bits should be extracted out of the 33 bits. A common and empirical approach to select the proper range is described in [ALT09]. According to this proposal, the filter design shall be exhaustively simulated with all possible input values. Then the effective bit range can be obtained from the simulation results. Indeed this method can prevent the implemented filter from saturating its outputs. However, if the coefficients are changed, the simulations should be made again. The input signals are sometimes unknown, and thus difficult to predict for an exhaustive simulation. Therefore, the aforementioned method is not suitable. Instead an iterative algorithm - as shown in Fig. 4.24 - is implemented.

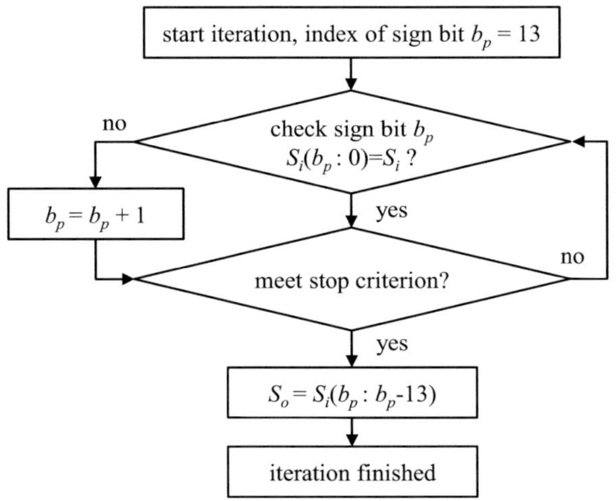

Fig. 4.24 Flowchart for iterative determination of effective bits.

Suppose S_i denotes the 33-bit signed filter output and S_o is the 14-bit signed DAC input. The variable b_p points to the most significant bit (MSB) of the desired effective range. It must refer to the index of the lowest sign bit of S_i. The other sign bits don't contribute to the dynamic range and should be discarded. At the beginning of the algorithm, as shown in Fig. 4.25, b_p is pointing to the 13th bit (the bit index begins with 0) of S_i and the least significant 14 bits of S_i are converted to a signed integer value and compared with S_i. An equality of both values indicates a valid value of b_p. Otherwise, b_P is incremented by one for the next comparison. Again, the 14 bits from index b_P to b_P -13 are converted to a signed integer value and compared with S_i. The process iterates until b_p does not change any more. The iteration can be stopped manually by the test operator any time. In our test practice, we first run a complete probing test cycle for the algorithm to determine the optimal effective bit range which is than kept constant for the rest of the time. The user can of course implement more sophisticated stop criteria, so that the algorithm can be terminated automatically. The resulting 14 bits are assigned to S_o and converted to analog signals. This algorithm can be applied whenever new filter coefficients are loaded into the emulator. In comparison with the simulation-based method, the real transmitted signals can be fed into the FIR filter directly. Therefore, the determination of the effective filter output is much faster, simpler and more realistic.

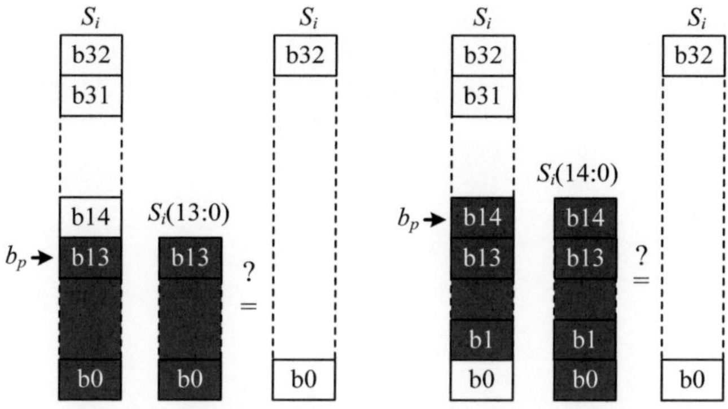

Fig. 4.25 Key steps of the proposed iterative algorithm.

4.4.7 Quantization Error

Generally speaking, three quantization effects are observed and can affect the performance of FIR filters: quantization of an analog signal performed during the analog to digital conversion (also known as AD noise), quantization of the filter coefficients and quantization of the filter operation results [CHA73]. These quantization effects and quantization errors are important practical aspects that must be considered for the implementation of digital filters. In the most popular quantization noise model a white noise source is associated to each rounding operation in the filter. Suppose the minimum quantization step is denoted by q. The mean and the variance of this white noise are zero and $q^2/12$ respectively. The noise samples are uniformly distributed between $-q/2$ and $q/2$. In addition, all noise sources are uncorrelated with each other and with the input signal [CHA732], [CHA73].

The AD noise determines the sampling accuracy. For a sinusoidal signal with maximum allowable amplitude, the achievable SNR in dB is dependent on the bit width B_i of the ADC, i.e.

$$SNR_{AD}(dB) = 1.76 + 6.02 \cdot B_i. \tag{4.62}$$

The ADC used in this thesis has 12-bit resolution, resulting in 74 dB. The AD noise at the filter input, denoted by e_{ADi}, is of zero-mean, and has a variance $q^2_{AD}/12$. The AD error at the filter output is

$$e_{ADo}(n) = \sum_{i=0}^{N_{fir}-1} e_{ADi}(i) \cdot h(n-i). \tag{4.63}$$

The variance of e_{ADo} is

$$\sigma^2_{AD} = \frac{q^2_{AD}}{12} \cdot \sum_{n=0}^{N_{fir}-1} h^2(n). \tag{4.64}$$

Independent of the coefficient scaling schemes of the filter in direct form, e_{ADo} always fulfills the relation

$$\sigma^2_{AD} \leq \frac{q^2_{AD}}{12}. \tag{4.65}$$

The error caused by quantizing the operation results within the filter - such as by the process discussed in 4.4.6 - is also called round-off error. Since this quantization is made at the end of the correlation, there is only one round-off error source. It is assumed that any overflow error is avoided by applying the aforementioned iterative algorithm. The influence of the round-off error is twofold: if the input signal is so small that the least 14 bits are sufficient for the filter output, and the upper 19 bits of the 33-bit word are the sign bits, then there is no degradation of accuracy; if the original output uses its 33rd bit as the sign bit and the remaining 32 bits as the data bits, the last 19 bits must be discarded. The round-off error is calculated similar to the AD noise and has the variance

$$\sigma^2_{\ rd} = \frac{q^2_{\ rd}}{12}.$$

(4.66)

Under consideration of both the AD noise and the round-off noise, the resulting SNR in linear scale is

$$SNR_{rd} = \frac{s^2_{\ M} \cdot \sum_{n=0}^{N_{fir}-1} |h(n)|^2}{\dfrac{q^2_{\ AD}}{12} \cdot \sum_{n=0}^{N_{fir}-1} |h(n)|^2 + \dfrac{q^2_{\ rd}}{12}},$$

(4.67)

and

$$q_{AD} = \frac{|s|_M}{2^{12}}.$$

(4.68)

With the help of the iterative algorithm, we get

$$\frac{|s|_M}{2} \leq \left| |s|_M \cdot \sum_{n=0}^{N_{fir}-1} h(n) \right| \leq |s|_M.$$

(4.69)

Therefore we have

$$\frac{q_{AD}}{8} \leq q_{rd} = \frac{\left| |s|_M \cdot \left| \sum_{n=0}^{N_{fir}-1} h(n) \right| \right|}{2^{14}} \leq \frac{q_{AD}}{4}.$$

(4.70)

Applying (4.70) in (4.67) we achieve

$$SNR_{AD} \cdot \frac{\sum_{n=0}^{N_{fir}-1} |h(n)|^2}{\sum_{n=0}^{N_{fir}-1} |h(n)|^2 + \frac{1}{16}} \leq SNR_{rd} \leq SNR_{AD} \cdot \frac{\sum_{n=0}^{N_{fir}-1} |h(n)|^2}{\sum_{n=0}^{N_{fir}-1} |h(n)|^2 + \frac{1}{64}}. \tag{4.71}$$

The final SNR is dependent on the quadratic sum of the filter coefficients. In the case of a single impulse, one gets

$$-0.26 \leq SNR_{rd}(dB) - SNR_{AD}(dB) \leq -0.07. \tag{4.72}$$

In the case where $h(0) = h(1) = \ldots = h(N_{fir}-1) = 1/N_{fir}$ and $N_{fir} = 1000$ we have

$$-18.02 \leq SNR_{rd}(dB) - SNR_{AD}(dB) \leq -12.21. \tag{4.73}$$

Obviously the degradation caused by the round-off error depends on the filter coefficients. In the worst case, the SNR can be reduced by 18 dB.

4.4.8 Loopback Test[2]

The quality of an emulated channel transfer function is not only influenced by the digital filter, but also by the AFE and coupling circuits. Fig. 4.26 illustrates the complete signal path from the transmitter (Tx) to the receiver (Rx). $x(t)$ and $y(t)$ are the signals at the Tx output and the Rx input respectively. $H_R(f)$ and $H_T(f)$ denote the transfer functions of the receiving and the transmitting paths respec-

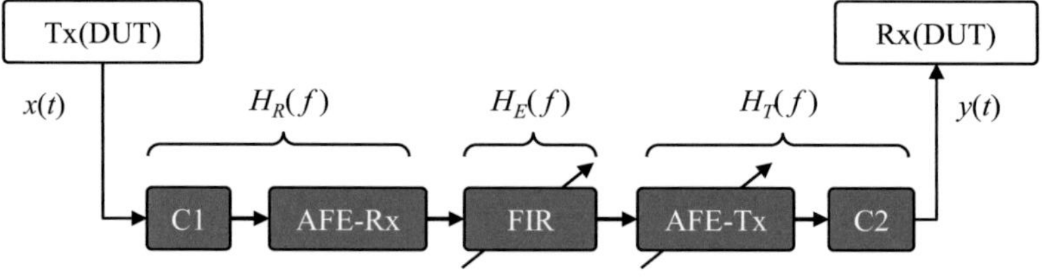

Fig. 4.26 Signal paths for the transmitted signal.

[2] This part has been published in [LIU11]: W. Liu, M. Sigle and K. Dostert, "Advanced Emulation of Channel Transfer Functions for Performance Evaluation of Powerline Modems," in Proc. IEEE Int. Symp. Power Line Commun. Applicat., Udine, Italy, Apr. 2011, pp. 446-451.

tively, either of which is mainly composed of a coupling circuit (C1 or C2) and an AFE. $H_E(f)$ is the transfer function of the digital filter. Since the AFE of the transmitting path (AFE-Tx) contains a variable gain amplifier (VGA), with which the stepwise attenuation of the transmitted signal is controlled, $H_T(f)$ may change during a performance test. The total transfer function $H(f)$ is

$$H(f) = H_R(f) \cdot H_E(f) \cdot H_T(f). \tag{4.74}$$

To achieve high accuracy it is necessary to check whether $H(f)$ meets all requirements or not. If not, $H_E(f)$ should be fine-tuned to reduce errors. For the tuning process, the actual $H(f)$ can be measured according to Fig. 4.27. The transmitting path is connected to the receiving path through a loopback cable. The DUT-Rx is kept attached to C2, so that its influence is also considered. A test signal $x'(t)$ is generated by the FPGA. The distorted counterpart $y'(t)$ is captured at the same time. The transfer function of the loopback circuit $H'(f)$ can be obtained by

$$H'(f) = Y'(f) / X'(f) = H_E(f) \cdot H_T(f) \cdot H_R(f), \tag{4.75}$$

where $X'(f)$ and $Y'(f)$ are the Fourier transforms of $x'(t)$ and $y'(t)$ respectively. Comparing (4.74) with (4.75) leads to the fact that the loopback transfer function $H'(f)$ is identical with the overall transfer function $H(f)$. Since $H'(f)$ can be obtained digitally, the whole process can be automated easily.

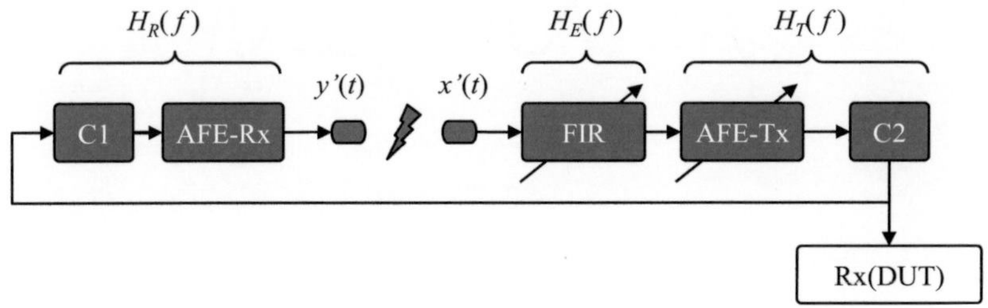

Fig. 4.27 Signal paths for a loopback test.

A chirp with a frequency span covering the whole band of interest is recommended as test signal. The achievable attenuation value A_{dB} is calculated by

$$A_{dB}(i) = 10 \cdot \log_{10}\left[\frac{E_y'(i)}{E_x'}\right], \tag{4.76}$$

where E_x' and E_y' are the energies of $x'(t)$ and $y'(t)$ respectively, and i is the corresponding control code.

4.4.9 Verification

The emulated CTF is measured with a vector network analyzer (VNA). Fig. 4.28 compares the desired CTF (desired), the designed CTF (sim), using the extended frequency sampling approach and the measured CTF (mea) of the hardware implementation. For the magnitude response, the overall quality of the emulation is high. There are deviations of the measured magnitude from the design in the frequency range 400 - 500 kHz. This is caused by the anti-aliasing filter which has a 6 dB-drop around 800 kHz. The measured group delay exhibits good agreement.

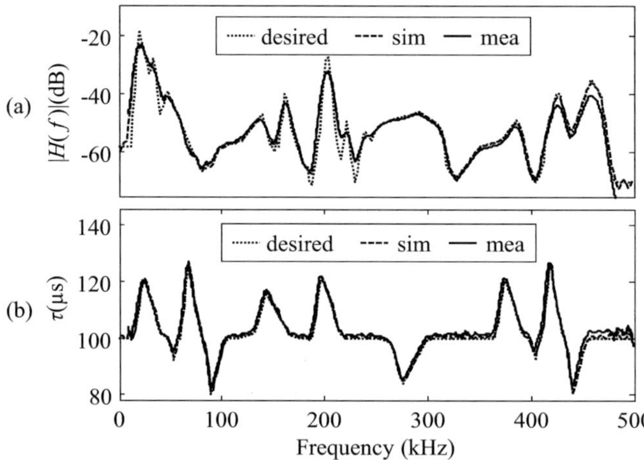

Fig. 4.28 Comparison of desired, simulated (sim) and measured (mea) magnitude (H) and group delay (τ) responses.

4.4.10 Realization of Time Variance

It has been mentioned that the channel response features both long- and short-term variances. This section focuses on the emulation of the latter. In [CAN08] the time-variance with 8 fundamental states per mains cycle is modeled. The transfer function in each state and the duration for the current state are used as the attributes of each state. Since fast convolution has been made in the frequency domain, each fundamental state should last for a multiple of FFT-frames. To improve the time-frame resolution, a linear interpolation is performed with the fundamental states. As a result, more frequency responses are created - one for each frame - and the resolution is improved in comparison with a single FFT-frame. With the help of the linear interpolation, the frequency response changes smoothly. The FFT-based fast convolution provides a convenient way for the linear interpolation in the frequency domain. A new frequency response can be calculated by simply adding an increment to the previous one at each frequency.

Since the FIR filter-based direct convolution is implemented in the time domain, it is necessary to realize the spectral interpolation using operations in the time domain too. Consider the spectral linear interpolation

$$H_n(k) = H_{n-1}(k) + \Delta H_m(k), \tag{4.77}$$

where n, m and k are indices of the interpolated state, fundamental states and frequency bin respectively. The increment $\Delta H_m(k)$ is obtained by

$$\Delta H_m(k) = \frac{H_{m+1}(k) - H_m(k)}{T_m}, \tag{4.78}$$

where T_m is the duration of the mth fundamental state. Obviously, $\Delta H_m(k)$ remains constant within a fundamental state. In the time domain, the impulse response $h_m(i)$ is obtained by

$$h_m(i) = \frac{1}{N} \cdot \sum_{k=0}^{N-1} H_m(k) \cdot e^{j \cdot \frac{2\pi \cdot k \cdot i}{N}}. \tag{4.79}$$

Inserting (4.77) into (4.79), the new impulse response can be expressed by the impulse response of the previous state and an increment, i.e.

$$h_n(i) = \frac{1}{N} \cdot \sum_{k=0}^{N-1} \left[H_{n-1}(k) + \Delta H_m(k) \right] \cdot e^{j \cdot \frac{2\pi \cdot k \cdot i}{N}} = h_{n-1}(i) + \Delta h_m(i), \tag{4.80}$$

where $\Delta h_m(i)$ is the inverse DFT of $\Delta H_m(k)$. Since $\Delta H_m(k)$ does not change within the mth state, $\Delta h_m(i)$ also remains unchanged. Fig. 4.29 shows frequency responses of 6 fundamental states, denoted by S_1 through S_6. Fig. 4.30 shows the spectra of the fundamental and the interpolated states.

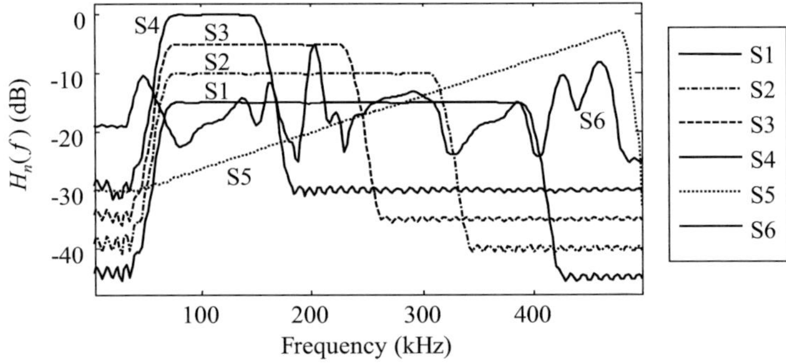

Fig. 4.29 Frequency responses of the fundamental states S1 through S6.

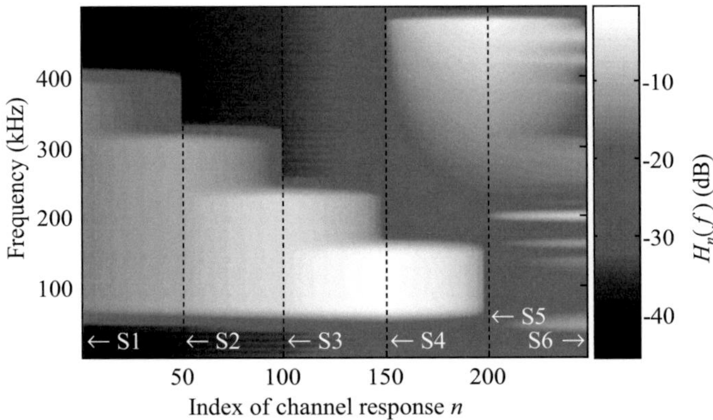

Fig. 4.30 Interpolated states of the frequency response.

These frequency responses are not taken from real-world measurements, but are created artificially to verify the interpolation. These fundamental states are interpolated linearly with 49 additional states between two consecutive fundamental states. The transmissions are smoothed and the boundaries are blurred. The direct convolution delivers the output for each new input at the sampling rate. Theoretically, the resolution can be as high as the sampling period, if the new filter coefficients are also available within each sampling period. Therefore, the resolution is only determined by the speed at which a complete coefficient set can be delivered to the FIR filter.

5 Noise Scenario

5.1 Typical Powerline Noise Scenarios

Noise plays a significant role in the channel impairment. Since a transmit signal is strongly attenuated in NB-PLC channel, the communication quality is determined to a high degree by the noise scenarios seen by the receiver. However, the noise scenario deviates from the traditional AWGN model. It has more complicated spectral characteristics and features time-varying behavior. Typical noise types in NB-PLC are colored background noise, impulsive noise and narrowband interferers [KAT06]. The noises are mostly generated by connected electrical appliances or by connection and disconnection of these devices. Typical noise sources are switching power supplies, light dimmers, silicon-controlled rectifiers, brush motors, monitors, and so on [FER10]. In many appliances, passive elements, such as ferrite cores for noise filtering, are found. Such components can generate or amplify periodic impulsive noise, due to their current-dependent nonlinear characteristics. This phenomenon has been analyzed in chapter 3.

Many appliances generate noise synchronously with the mains voltage. The resulting noise exhibits periodic features which are synchronized to the mains frequency. In addition, the channel attenuation between a noise source and the measurement location can also change as a result of time-varying access impedance and/or network topology. Thus, this kind of channel fluctuations, which are not caused by the noise itself, will also lead to time-variant noise behavior [KAT06]. The noise level and the PSD are determined by the noise source and the equivalent access impedance seen by the noise source. Since the access impedance is frequency-selective instead of being flat, the resulting noise spectrum is not white. Instead, the noise has a complicated frequency-selective PSD.

5.2 Modeling Powerline Noise: State-of-the-Art

Much effort has been made to model powerline noise scenarios. One popular model is the probability density function (PDF) for the non-Gaussian noise, proposed by Middleton [MID77]. The PDF of impulsive noise is expressed as a sum of Gaussian functions with different variances. The advantage of this model is a convenient expression of impulsive noise types by using a function with a reduced number of parameters. However, this approach is unable to describe time-domain courses of the noise.

In order to add the time-domain information, [KAT06] has proposed an extended model which describes the temporal behavior of the noise by changing the variance of a Gaussian process periodically. The fluctuation of the variance is synchronized with the mains voltage, and the period corresponds to one half of the mains cycle. A following processing of the noise within a digital filter shapes the colored noise spectrum. Although this model can control the noise waveform in the time domain and the noise has a well-defined colored power spectrum, it is unable to model time-varying PSDs. In order to capture the cyclo-stationary features in both the time and frequency domain, [MAR12] has proposed to use a spectro-temporal cyclo-stationary model. This model first divides a noise waveform into several regions within which the noise is considered to be stationary. The instantaneous noise PSD in each region is obtained and reproduced by using a digital shaping filter. The modeled artificial noise is generated by processing AWGN with spectral shaping filters. By constructing a periodic and time-variant filter bank with linear time-invariant spectral shaping filters, this model is able to combine the time-variance and the spectral contents. However, it is insufficient to model the noise source by solely using AWGN. There are periodic impulsive noise types with an exponentially decay in their waveforms and a regular frequency dependency in their phase vector. They originate from high-energy pulses with extremely narrow pulse widths. The PSD and the phase vector calculated at each frequency reflect the magnitude and phase responses of the channel. An example of such a noise waveform and its spectral characteristics can be found in Fig. 5.29 and Fig. 5.30 later in this chapter. This kind of impulsive noise can hardly be modeled with an AWGN-based model. In addition, the spectro-temporal cyclo-stationary model needs a large number of filter coefficients. For example, to reach a resolution of 1 kHz at a sampling rate of 2 MSps, a digital filter needs 2000 filter coefficients. If 5 regions are identified within a mains cycle, 10000 coefficients will be needed for modeling the complete noise scenario. Moreover, there are noise types that are not stationary and their appearances make the clear definition of stationary regions impossible. They cannot be reproduced by using a simple filter bank with a limited number of digital filters. Most noise sources in powerline networks are difficult to identify. In addition, different electrical appliances and power consumption activities create a variety of noise with different spectral content and different time behavior. Thus, it is a particular challenge to provide an according noise scenario with a universal model. Toward an acceptable solution, it is necessary to classify the noise types first, and then to model different noise types in different ways.

5.3 Additive White Gaussian Noise (AWGN)[3]

Channels with flat attenuation and AWGN are widely used for investigating communication environments. Although AWGN alone is not sufficient for describing PLC channels, the Gaussian process is useful in several steps, e.g. to generate background noise. It is also an important starting point to obtain more complicated random processes. This section presents the emulation of Gaussian random numbers by pure digital technology. The PDF of a Gaussian distribution is

$$f(x) = \frac{1}{\sqrt{2\pi\sigma^2}} e^{-\frac{(x-\mu)^2}{2\sigma^2}}, \tag{5.1}$$

where μ is the mean value and σ is the standard deviation. The rapid generation of high quality Gaussian random numbers for computer-based simulations has been well studied for years. A plenty of algorithms has been developed, for example, the central limit theorem (CLT) method, the Ziggurat method, the Box-Muller method, and others – see e.g. [THO07] [BOX58] [MAR64] and [DEV86]. Some methods have been implemented into modern programmable devices, such as FPGAs – see e.g. in [FAN04]. In [FAN08] a noise generator has been proposed within a BER tester for digital communication interfaces. Most of the work has made effort, either to speed up the generation of the random numbers, or to improve the accuracy of the "tail" areas of the PDFs. In practice, where the generated noise is used to test real communication interfaces or systems, the digital

---

[3] Most parts of this section are direct copy of the publication [LIU112]: W. Liu, C. Li and K. Dostert, "Emulation of AWGN for noise margin test of powerline communication systems," in Proc. IEEE Int. Symp. Power Line Commun. Applicat., Udine, Italy, Apr. 2011, pp. 225-230.

noise samples must be converted into analog waveforms by DACs. The output of such a DAC must be filtered with a reconstruction lowpass filter. Then amplification is needed to extend the waveform to the desired dynamic range. Furthermore, in context with powerline data transmission channels, the noise generation system and the DUTs are working under mains voltage. Thus, coupling circuits must be used to separate the components of the noise generator from the mains voltage. The quality of the emulated noise appearing at the input of a DUT can be degraded if the whole processing path is not handled properly. Therefore, it is necessary to consider the influence of the total processing path on the characteristics of the generated noise.

5.3.1 Overview of Existing Methods

This section introduces four popular methods to generate Gaussian random numbers: the CLT method, the Box-Muller method, the polar method and the direct rejection method [THO07].

Central-Limit-Theorem based Method

The CLT approach is based on the theorem that the mean of a large number of independent identically distributed (i.i.d.) random variables will have a Gaussian distribution [RIC95]. The implementation sums up dozens of uniformly distributed random numbers and outputs the sum as the desired noise. This method appears extremely simple. However, the convergence toward the wanted Gaussian PDF is very slow [FAN04].

Box-Muller Method

The Box-Muller method generates pairs of Gaussian random numbers that represent the coordinates on a two-dimensional plane. The magnitude and the phase can be generated from uniform random numbers. Let M_1 and M_2 be independent random variables that are uniformly distributed in the interval (0, 1]. The random variables Z_1 and Z_2 obtained by transforming M_1 and M_2 into the coordinates

$$Z_1 = \sqrt{-2\ln(M_1)} \cdot \cos(2\pi M_2) \tag{5.2}$$

and

$$Z_2 = \sqrt{-2\ln\left(M_1\right)} \cdot \sin\left(2\pi M_2\right) \tag{5.3}$$

form a pair of independent random variables of a Gaussian distribution [BOX58].

Polar Method

In the polar method, we use the two independent uniform random variables M_1 and M_2, as well as the sum $s = M_1^2 + M_2^2$. Then the two new random variables

$$Z_1 = \frac{\sqrt{-2\ln(s)} \cdot M_1}{\sqrt{s}} \tag{5.4}$$

and

$$Z_2 = \frac{\sqrt{-2\ln(s)} \cdot M_2}{\sqrt{s}} \tag{5.5}$$

represent two independent Gaussian distributed variables [MAR64]. The polar method can be seen as an improvement of the Box-Muller method. It eliminates the trigonometric calculations that are usually slow and take more memory resources.

Rejection Method

The rejection method is based on a fundamental theorem of densities described in [DEV86]: if X is a random vector with the density function $f(X)$, and M is an independent uniform variable within [0, 1], then the vector pair $(X, cMf(X),)$ is uniformly distributed on the area A that is covered by X and the curve $cf(X)$, where c is a positive constant. Vice versa, if a random pair (X, M) is uniformly distributed on an area covered by X and the curve $c f(X)$, if M is uniformly distributed within the interval $[0, cf(X)]$, then $f(X)$ must be the density function of X. Based upon this theorem we obtain the random vector with desired density function f in following steps:

- Generate two independent uniform random vectors M_1 and M_2

- Calculate the function value $cf(M_1)$

- Compare M_2 with $cf(M_1)$, if $M_2 < cf(M_1)$, accept M_2 as a valid element of the desired random vector, otherwise reject it and repeat from the first step

The accepted value X has the density function $cf(X)$. c is a scaling factor used to make sure that the integral over the density function is equal to one. The rejection method doesn't need sophisticated calculations. It can even be used to generate arbitrary density functions which cannot be obtained by other methods. However, it must also be noted that more than just one pair of random vectors are used to generate a single valid random number. Thus, the efficiency is limited, and it is not suitable for high-speed applications.

5.3.2 Hardware Implementations

Generation of Uniformly Distributed Random Numbers

Before we generate white Gaussian noise, we should generate uniformly distributed random numbers first. A maximum length sequence (MLS), also called M-sequence is usually used to generate pseudo-random numbers. The necessary operations are performed by maximal linear feedback shift registers (LFSR). Two methods are introduced in [GOE04] to construct a M-sequence. The first method utilizes only one LFSR out of which M taps are combined to form an M-bit random number. The other one uses M LFSRs that operate in parallel, but with different seeds. The output of each LFSR is parallelized and used to construct the desired M-sequence. In this thesis, the second method is employed.

Implementing the Box-Muller Method

The block diagram of a FPGA-implementation of the Box-Muller method is shown in Fig. 5.1. The LFSRs blocks produce two 12-bit uniform random numbers M_1 and M_2. All necessary samples of the trigonometric functions and the root of the natural logarithm are stored in the ROM blocks as look-up tables. The application of look-up tables eliminates complicated calculations. The two outputs Z_1 and Z_2 are obtained by multiplying W with P and Q respectively. Since all the values of M_1 and M_2 can be used to generate the Gaussian random numbers, and each pair of inputs results in a valid pair of outputs, the throughput is the same as the generation rate of M_1 and M_2. Thus, all of the functional blocks are driven by the same clock signal.

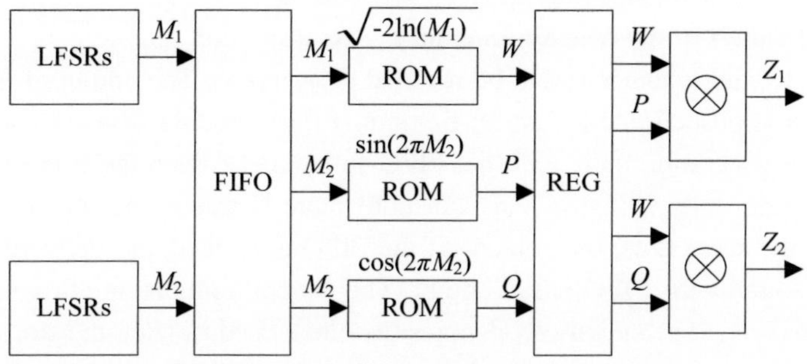

Fig. 5.1 Block diagram for the Box-Muller method.

Implementing the Polar Method

The implementation of the polar method is similar to that of Box-Muller method. The block diagram is shown in Fig. 5.2. Since the trigonometric functions are not involved, only one look-up table is needed. The functional block CMB combines M_1 and M_2 together to form the address for the look-up table. The clamp units are used to match the bit width of U and V to the inputs of the multipliers.

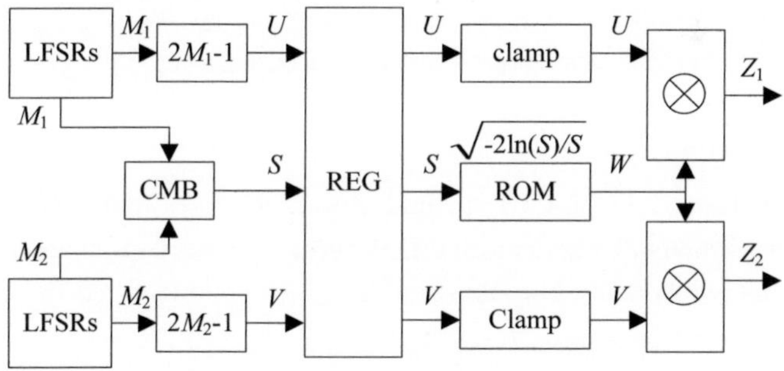

Fig. 5.2 Block diagram for the polar method.

Implementing the Direct Rejection Method

The implementation of the direct rejection method is relatively simple. Similar to the previous designs, two uniform random vectors M_1 and M_2 are created using

LFSRs. M_1 is scaled and shifted with an offset to access the ROM which contains the scaled values of the density function f. The arbitration unit compares V with $f(U)$ and decides whether V will be rejected or accepted. The output of the arbitration unit is pushed into a first-in, first-out (FIFO) module whose status is fed back to the generation units with the FIFO-full signal. When the FIFO is filled, the generation units will stop and wait until there is free space for new values again. At the same time, the content of the FIFO is fetched and delivered to the DAC, as soon as the FIFO is not empty. The generation units are driven by the clock signal clk_1. The signal clk_2 denotes the clock used to fetch the stored numbers and to drive the DAC. To make sure that the noise generation will not be broken by lack of random numbers, the frequency of clk_1 is made 5 times higher than that of clk_2.

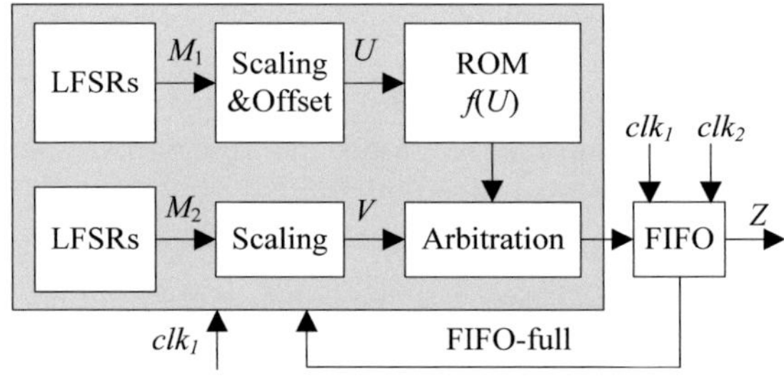

Fig. 5.3 Block diagram for the direct rejection method.

Resource Consumption

As shown in Table 5.1, the CLT method utilizes the most logic cells, however, neither internal memory nor dedicated DSP blocks are needed. In addition, it has the least clock redundancy. The Box-Muller and the polar methods use fewer log-

Table 5.1 Hardware Resource Consumption for the above Methods

	Logic cells	M9Ks	DSP 16x16	Redundancy
CLT	2692	0	0	3 clocks
Box-Muller	358	2	1	5 clocks
Polar	384	1	2	4 clocks
Direct Rejection	688	15	0	-

ic cells. However, dedicated memory and DSP blocks are necessary. The direct rejection method needs a large number of memory blocks. The clock redundancy cannot be determined exactly.

5.3.3 Influence of Analog Components

The generated Gaussian random numbers cannot be applied to a DUT directly. A conversion to analog waveforms by a DAC is necessary. The DAC output should be smoothed by a reconstruction low pass filter, and the signal has to pass the coupling circuitry before it finally arrives at the DUT. This section discusses the influence of the analog components on the distribution and the spectral characteristics of the generated white Gaussian noise.

5.3.3.1 Impact of the Linear Filter

We can consider the overall noise path to be a system having Gaussian random numbers as an input and the waveform measured at the input of the DUT as an output. Since the circuitry is designed in such a way that it applies linear filtering to the waveform only, it is reasonable to treat this system as linear. The output of the system $y(t)$ can be obtained by convoluting the input signal $x(t)$ with the system's impulse response $h(t)$, i.e.

$$y(t) = h(t) * x(t). \tag{5.6}$$

Apparently the system is not memory-less. The statistic properties of the current output are determined not only by the current input, but by a sequence of words of the input vector. Generally speaking, it is impossible to obtain a complete specification of the output process by using a closed form expression of the input process. Fortunately, the Gaussian process, as an exception, can be completely described by its mean and variance. Then, the output process of a linear filter, driven by a Gaussian random process, will also be Gaussian. According to [GRA01], the mean function $m_y(t)$ and the covariance function $K_y(t,s)$ of a weakly stationary input process can be obtained by

$$m_y(t) = m_x \int h(t)dt \tag{5.7}$$

and

$$K_y(t,s) = \int dm \int dn K_x\big[(t-s)-(m-n)\big]h(m)h(n), \tag{5.8}$$

where m_x and $K_x(t,s)$ are the mean and covariance functions of the input process respectively. The power spectral density (PSD) of the filter output $S_{yy}(f)$ can be obtained by

$$S_{yy}(f) = |H(f)|^2 \cdot S_{xx}(f), \tag{5.9}$$

where $S_{xx}(f)$ denotes the PSD of the filter input, and $H(f)$ is the frequency response of the filter [HAE01].

5.3.3.2 Influence of Amplification

If an operational amplifier has a gain factor a and an offset b, then the PDF of its output $f_y(y)$ is

$$f_y(y) = \frac{1}{|a|} f_x\left(\frac{y-b}{a}\right), \tag{5.10}$$

where $f_x(x)$ is the PDF of the input signal [HAE01].

5.3.3.3 Impact of Non-Linear Distortion

Non-linear distortion may be caused by the saturation of a coupling transformer or clipping by a surge-protection diode. The effect can be considered as memoryless. Clipping can be described by

$$y = \begin{cases} x, |x| < c \\ c, |x| \geq c \end{cases}, \tag{5.11}$$

where c is the threshold voltage of the surge-protection diode. The PDF of y is therefore

$$f_y(y) = \begin{cases} f_x(y), |y| < c \\ A\delta(y+c) + A\delta(y-c), |y| \geq c \end{cases}, \tag{5.12}$$

where $\delta(y)$ is a Dirac impulse, and A is the probability that x is larger than c. By applying (5.1) we have

$$A = \mathrm{Prob}\{x > c\} = \frac{1}{2}\left[1 - \mathrm{erf}\left(\frac{c}{\sqrt{2\sigma_x^2}}\right)\right],\tag{5.13}$$

where σ_x^2 denotes the signal power [BAN02].

5.3.4 Evaluation

5.3.4.1 Evaluation Platform

In order to compare the qualities of different implementations and study the influence of the analog circuitry, an evaluation platform is constructed, as shown in Fig. 5.4. The PC interface has a buffer of the size 65536. As soon as the buffer is filled, a data transaction takes place, and the samples are transferred to a PC over an USB cable. Since the data transfer is much faster than the generation of the random numbers, it is guaranteed that the PC can obtain all samples for verification. The waveform of n_2 at the output of the coupling circuit is observed and recorded by a PC-based oscilloscope. Both the histogram and the PSD of n_1 and n_2 are used for further investigation.

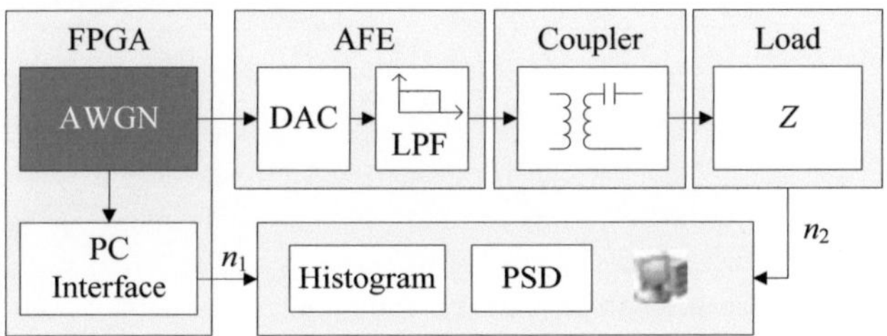

Fig. 5.4 Setup to verify the qualities of different implementations.

5.3.4.2 Histogram

In order to evaluate the distributions of n_1 and n_2, histograms are constructed with regular-spaced bins (segmentation of all possible random numbers). The number of samples falling into each bin is normalized and compared with the theoretical

PDF of a Gaussian distribution. The red and blue curves in the histogram plots in the following figures are the normalized histograms, and the Gaussian distributed PDFs respectively. The abscissa refers to the normalized bins ranging from -1 to 1. The ordinate shows the probability density. For the convenience of plotting, both the x- and y-labels are removed. Fig. 5.5 shows the histograms of the digital value n_1 generated by means of the aforementioned methods. All of them exhibit good agreement with their corresponding theoretical PDFs. The peaks around bin 0 for the polar and the Box-Muller methods result from an unfavorable arrangement of the bin boundaries.

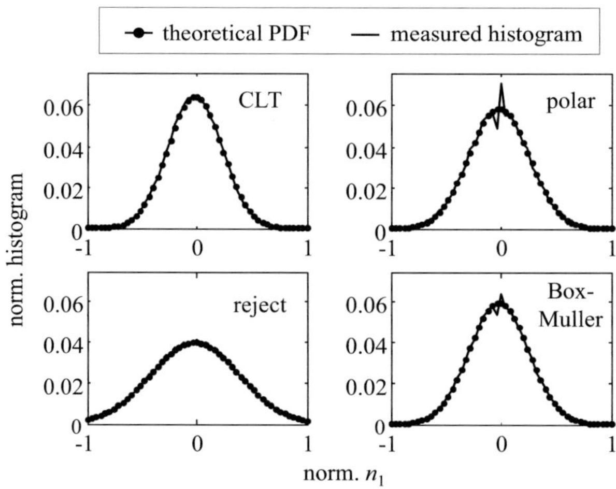

Fig. 5.5 Normalized histograms for n_1.

Table 5.2 Measured Mean Values μ and Standard Deviations σ for n_1

	μ	σ
CLT	-0.018	0.248
Polar	-0.021	0.271
Rejection	-0.019	0.405
Box-Muller	-0.021	0.268

Table 5.2 lists the measured mean values and the standard deviations of the four distributions. In the implementations using the polar, the reject and the Box-Muller methods, both values are specified and controlled by the look-up tables stored in the ROMs. In the design based on the CLT method, these values can also be modified by adding offset and scaling factor to the uniformly distributed random numbers. However, some effort should be spent to achieve the desired values. From this point of view, the CLT approach has less flexibility, but is simple to implement.

The measured distributions of n_2 are shown in Fig. 5.6 and listed in Table 5.3. The CLT, the polar and the Box-Muller methods provide better results than as the

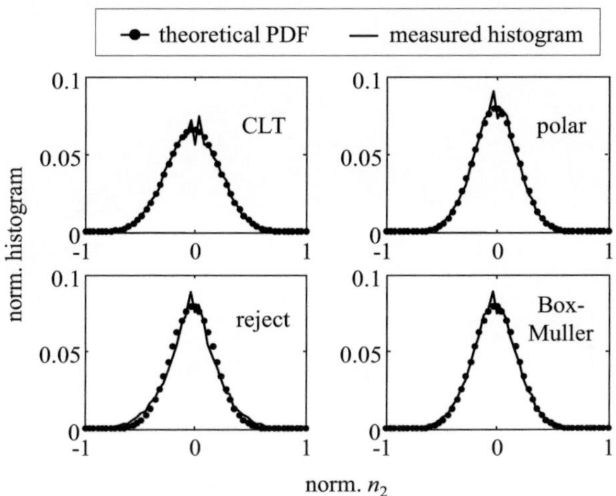

Fig. 5.6 Normalized histograms for n_2.

Table 5.3 Measured Mean Values μ and Standard Deviations σ for n_2

	μ	σ
CLT	-0.020	0.240
Polar	-0.020	0.200
Rejection	-0.020	0.200
Box-Muller	-0.020	0.200

direct rejection method. The degradation of quality will be explained by means of the PDF and the spectrum later. The distributions have the same non-zero mean value. However, this value is not resulting from n_2 itself, but from the offset of the oscilloscope, because the coupler exhibits high-pass characteristics, so that a DC component of n_2 cannot be transferred by the coupler. The standard deviation is controlled by setting the gain factor of the variable gain amplifier.

5.3.4.3 Power Spectral Density

Besides the Gaussian distribution, a constant power spectral density is the second feature characterizing white Gaussian noise. Fig. 5.7 shows the measured PSD curves of n_2. All curves have band-pass characteristics. The PSDs corresponding

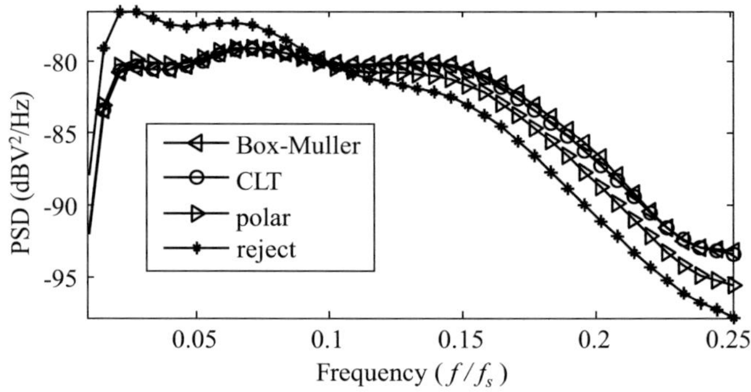

Fig. 5.7 PSD of n_2 without load.

to the CLT, the polar and the Box-Muller methods have similar courses with sharp bounds at the lower cut-off frequency, slight fluctuations in the pass band and a gentle transition from pass-band to stop-band towards high frequencies. This agrees with the transfer function of the signal path. The PSD curve referring to the rejection method lies above the others in the low frequency range, and below them in the high frequency range. This frequency selective attenuation indicates a loss of high frequency components during the DA conversion. During the verification process, it has been observed that the FIFO shown in Fig. 5.3 became empty sometimes. Apparently the speed of noise generation was not high enough. When there was no new output available, the same value was reproduced. Therefore, the quality degraded, and the PSD suffered from loss at high frequencies. To overcome this problem, the frequency of clk_1 should be further enhanced.

5.3.4.4 Influence of Loads on the Noise Characteristics

To investigate the influence of different loads on the characteristics of the generated white Gaussian noise, four different impedances Z0 through Z3 are connected to the emulator one after the other. The noise is generated by utilizing the Box-Muller method.

Table 5.4 Different Loads for Testing

Loads	Description
Z0	Input impedance of oscilloscope: 1 MΩ
Z1	LC serial circuit. L = 380 μH, C=470 nF
Z2	RC serial circuit. R = 5 Ω, C = 1 μF
Z3	RC serial circuit. R = 1 Ω, C = 1 μF

The PSDs of the noise measured across the loads are shown in Fig. 5.8. Since Z0 is extremely large, it doesn't change the spectrum. The other three loads introduce frequency-dependent impairments to the original spectrum. The spectra are no longer flat over the frequency range of interest, and the noise is no longer 'white'.

5.3.4.5 Impact of Amplification and Clipping

To study the impact of an amplifier and the non-linear distortion (clipping) caused by a suppressor diode, the power level of the noise is increased step by step, until the peaks of the noise waveform are partly cut off by the suppressor

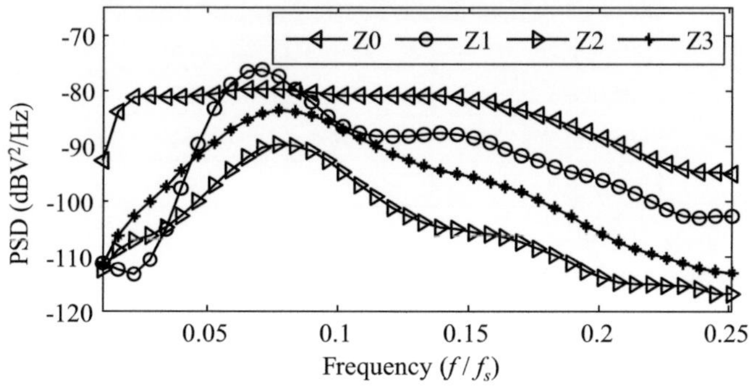

Fig. 5.8 PSDs of n_2 with different loads.

diode at the output of the emulator. The measured PDFs and the mean and deviation are shown in Fig. 5.9 and Table 5.5 respectively. Apparently the tail area around bin value 1 and -1 are destroyed by the clipping effect. To prevent the artificial noise from being clipped, care should be taken when selecting the suppressor diodes. The limiting voltage of these diodes should always be slightly larger than the input range of the receiver under test.

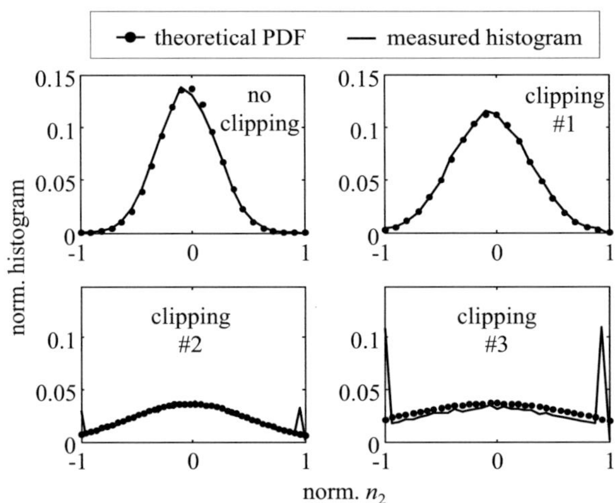

Fig. 5.9 PDFs of n_2 during clipping tests.

Table 5.5 Parameters of Noise Measured in Clipping Tests

	μ	σ
no clipping	-0.04	0.26
clipping 1	-0.05	0.35
clipping 2	-0.03	0.54
clipping 3	-0.03	0.93

5.4 Narrowband Interferers

This kind of interference features significant peak noise levels in the frequency domain in comparison with the background noise. In broadband PLC, this noise class exhibits either multiple equidistant discrete frequency components or a single frequency. Their envelopes may remain time-invariant over long periods, or may change periodically and be synchronous with the mains voltage [COR10]. In NB-PLC, disturbances at the frequencies 25, 30, 49, 55, 75 and 82 kHz have been measured in real-world powerline environments. They could be probably caused by switching power supplies. It has also been reported that narrowband noise appears preferably at frequencies below 140 kHz or above 410 kHz. The average bandwidth is about 3 kHz [BAU06]. The narrowband interferers with time-invariant amplitude envelopes have been well investigated. Modeling and emulating this kind of noise is also straightforward. Due to the time-varying nature of the powerline network, the envelopes of a number of narrowband interferers also exhibit dynamic behavior. The following sections introduce an approach for the estimation of such time-varying envelopes. A simplified model will be proposed for describing these envelopes. Finally, a phase-accumulation method will be introduced for the emulation of sinusoidal waveforms of narrowband noise.

5.4.1 Estimation of Noise Powers

Fig. 5.10 (a) shows a noise segment acquired in a university laboratory. This segment lasts for 40 ms, and the waveform in the time domain is dominated by impulsive components. Furthermore, the overall envelope of the non-impulsive

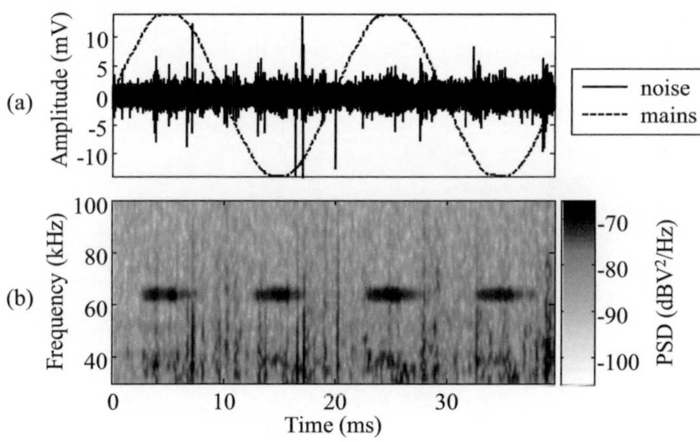

Fig. 5.10 Narrowband noise with cyclo-stationary feature. (a) Waveforms of noise and scaled mains voltage in the time domain, (b) STFT of the noise (Blackman window, window length: 500 µs, overlap ratio: 83.33%). Both (a) and (b) share the same abscissa.

components changes with time. Generally speaking, the location and the bandwidth of a narrowband interferer can be determined intuitively with the help of STFT by considering the aforementioned narrow and high-level peak in the frequency domain and the continuity or periodicity in the time domain. As shown in the spectrogram of the STFT in plot (b), there is a significant spectral component at around 64 kHz with a bandwidth of 4 kHz. The spectral density changes periodically. The local maxima are synchronous with the peaks of the mains voltage. It is a good example of narrowband interference with cyclo-stationary features. In order to estimate the time-dependence of the narrowband noise envelope, the influence of the other noise types, such as colored background noise and impulsive noise should be reduced as much as possible.

Fig. 5.11 shows the STFT of the noise segment from another point of view. The axis ticks as well as the color bar are ignored to simplify the illustration. b_M denotes the frequency band which contains the most power of the narrowband interferer, in this example, it is located between 62 and 66 kHz. b_L and b_R are the neighboring bands to the left and right of b_M respectively. b_L and b_R have the same bandwidths as b_M. Gaps are inserted between b_L and b_M as well as between b_M and b_R, so that the tails of each band will not interfere the other two bands.

The PSD of the background noise is relatively small compared to that of the narrowband noise. Therefore, for simplicity it is assumed that the background noise has the same noise power spectral density in all three bands. The spectra of

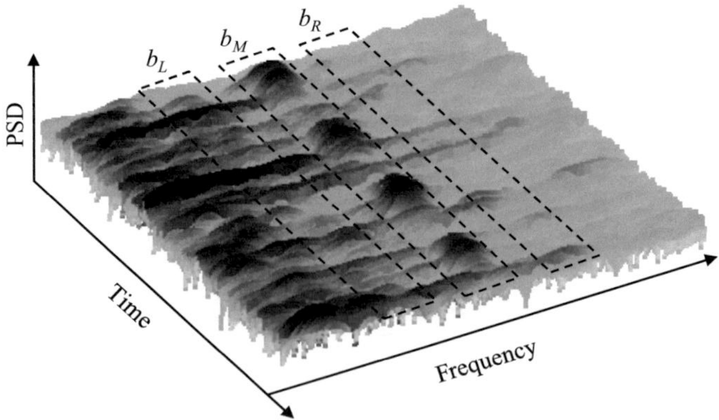

Fig. 5.11 STFT of a noise segment. b_M is the center frequency band. b_L and b_R are the left and right neighboring bands respectively.

short impulses usually exhibit wideband character. The PSD values are decreasing over frequency. Since these three bands are relatively short in comparison with the bandwidth of the impulsive noise, and they are close to each other, it is assumed that the power of the impulsive noise is a linear function of the frequency. Therefore the power falling into b_M is approximated by the mean value of the noise powers in b_L and b_R for this impulsive noise, i.e.

$$P_M(t)\big|_{dB} = \frac{P_L(t)\big|_{dB} + P_R(t)\big|_{dB}}{2},\tag{5.14}$$

where $P_M(t)|_{dB}$, $P_L(t)|_{dB}$ and $P_R(t)|_{dB}$ denote the powers of impulsive noise in a logarithmic scale within b_M, b_L and b_R respectively. The power of the impulsive noise in b_M can be estimated using

$$P_M = 10^{[\log_{10}(P_L)+\log_{10}(P_R)]/2}.\tag{5.15}$$

5.4.2 Reconstruction of Noise Waveforms

Fig. 5.12 shows a flow chart for the reconstruction of the narrowband interferer $n_{nbn}(t)$. The basic idea is to estimate the envelope and the oscillation waveform

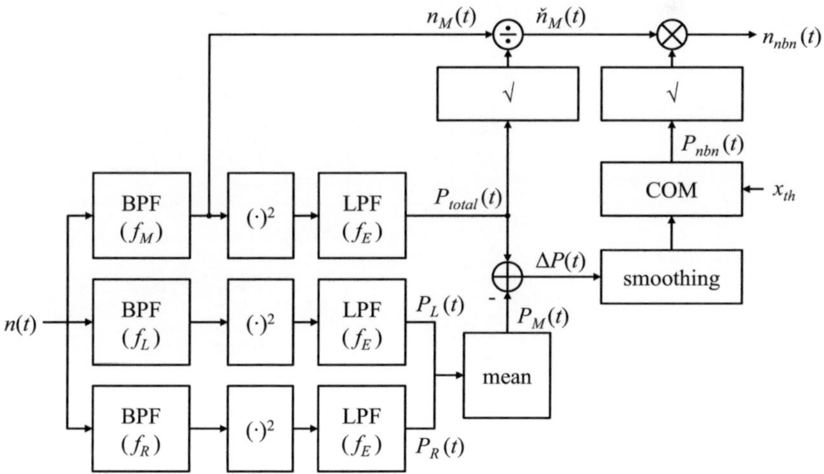

Fig. 5.12 Flow chart for detecting and extracting a single-frequency narrowband interferer.

$\check{n}_M(t)$ separately and then modulate $\check{n}_M(t)$ using $A_M(t)$. Three FIR bandpass filters are used to obtain the waveforms of the noise components located in each band. f_M, f_L and f_R denote the center frequencies of b_M, b_L and b_R respectively. Each filter is followed by a squaring operation and a lowpass filter with cutoff frequency f_E. These two components are used to estimate the time-variant envelope of the total PSD falling into each frequency band. $P_{total}(t)$, $P_L(t)$ and $P_R(t)$ denote the envelopes for the center, the left and the right band respectively. Fig. 5.13 (a) shows an example of the estimated envelopes for $P_L(t)$, $P_{total}(t)$ and $P_R(t)$ respectively. The zones A_1 to A_5 cover impulses that superimpose the narrowband interferer, while B_1 and B_2 cover impulses appearing in the intervals between two narrowband interferers. $n_M(t)$ in Fig. 5.13 (b) is the filtered waveform corresponding to the narrowband interferer superimposed by impulsive and background noise. $P_L(t)$ and $P_R(t)$ are used to estimate the non-narrowband noise power $P_M(t)$ for the center band.

In a next step, $P_M(t)$ is subtracted from $P_{total}(t)$ and the difference $\Delta P(t)$ contains mainly the narrowband interferer. The error caused by the background noise can be further reduced by smoothing $\Delta P(t)$ properly. A comparator compares the smoothed $\Delta P(t)$ with a threshold x_{th}. A value of $\Delta P(t)$ greater than x_{th} indicates a valid narrowband noise envelope. Otherwise $\Delta P(t)$ is forced to 0. In this way, the

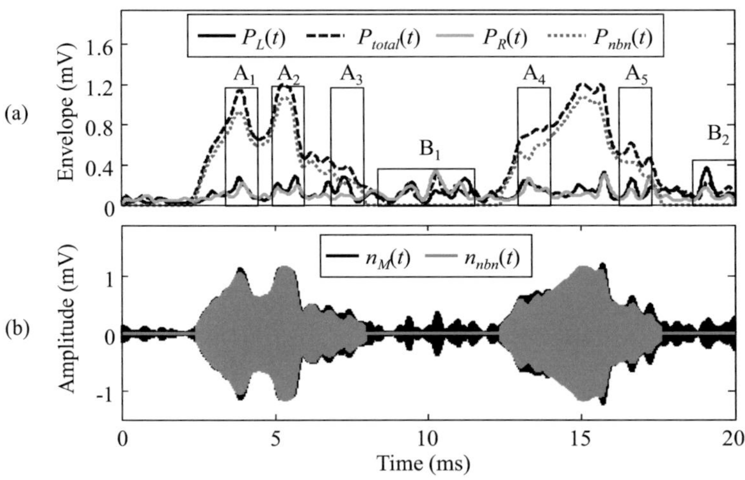

Fig. 5.13 (a) Estimating envelopes for all three sub-bands, (b) extracting the envelope of the narrowband interferer. Both (a) and (b) share the same abscissa.

power envelope of the narrowband interferer $P_{nbn}(t)$ can be estimated. Its square root $A_M(t)$ is the expected envelope of $n_{nbn}(t)$. The key aspect in the extraction of $\check{n}_M(t)$ is to keep $\check{n}_M(t)$ in phase with $n_M(t)$. The first idea is to estimate the phase of $n_M(t)$ and use it to synthesize a sinusoidal waveform. The phase estimation algorithm can be very simple if the frequency doesn't change over the time, otherwise a sophisticated frequency tracking strategy must be implemented. An alternative method is to estimate and compensate the fluctuation in the envelope of $n_M(t)$. The spectral feature is not affected in this way, therefore it can also be applied even if $n_M(t)$ exhibits a time-varying frequency. This method is implemented by simply dividing $n_M(t)$ by $A_{total}(t)$ which denotes the square root of $P_{total}(t)$.

In the last step, $n_{nbn}(t)$ is obtained by multiplying $\check{n}_M(t)$ by $A_M(t)$. After having estimated the envelope and synthesized the narrowband interferer, the synthesized waveform $n_{nbn}(t)$ is removed from the original noise $n(t)$ in the time domain. Fig. 5.14 shows the remaining noise waveform in plot (a) and the spectrogram of the STFT in plot (b). In comparison with Fig. 5.10, the investigated narrowband interferer disappears from the noise segment, while the background noise and the impulsive noise are not affected.

5.4.3 Modeling Narrowband Noise Envelopes

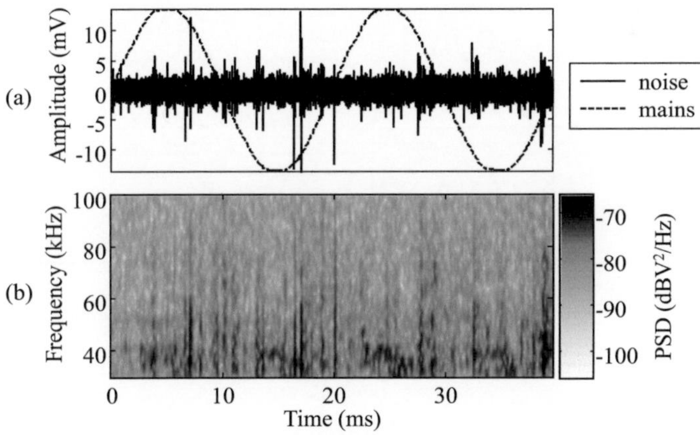

Fig. 5.14 Waveform and STFT of residual noise, the cyclo-stationary narrowband disturbance has been removed. (a) Waveforms of modified noise and scaled mains voltage in the time domain, (b) STFT of modified noise (Blackman window, window length: 500 µs, overlap ratio: 83.33%). Both (a) and (b) share the same abscissa.

The individual envelope of narrowband noise can be modeled by one or more un-symmetrical triangular functions. The shape and the location of a normalized peak is determined by

$$
y(t) = \begin{cases}
0, & t \leq t_1 \\[2mm]
\dfrac{t - t_1}{t_2 - t_1}, & t_1 \leq t < t_2 \\[4mm]
& \\[2mm]
\dfrac{t_3 - t}{t_3 - t_2}, & t_2 \leq t < t_3 \\[4mm]
0, & t_3 \leq t
\end{cases}
\tag{5.16}
$$

where t_1, t_2 and t_3 are the time instants for begin, peak and end of a triangular curve respectively. This kind of curves can be described by a triangular-shaped built-in membership function in Matlab [MAT01]. Fig. 5.15 shows a normalized measured envelope and the reconstructed artificial envelope in plot (a). The simulation is a sum of three fundamental shapes $y_1(t)$, $y_2(t)$ and $y_3(t)$.

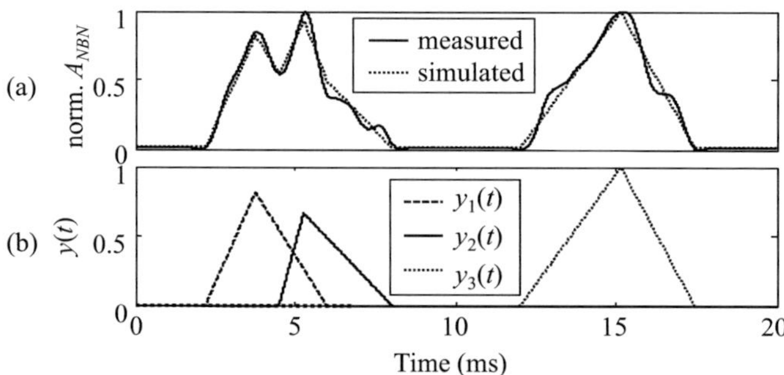

Fig. 5.15 Measured and simulated envelopes of narrowband interferers.

5.4.4 Emulating Narrowband Interferers

The emulation of narrow band noise covers noise types with sinusoidal waveform. According to [BAU05], this kind of noise has a bandwidth of several kHz. The amplitude level is greater than that of background noise. A narrow band noise model features three parameters: the center frequency f_m, the bandwidth Δf, and the amplitude A_{NBN}. The center frequency refers to the fundamental oscillation of the noise. The center frequency may change permanently, and its variation range is indicated by the bandwidth. The amplitude is a measure of the noise voltage level. The generation of a sinusoidal waveform is based on the phase accumulation method. As shown in Fig. 5.16, this method utilizes a phase accumulator and a look-up table. The phase accumulator consists of an adder and a phase register. The current phase $P(n+1)$ results from the modulo-N sum of the previous phase $P(n)$ with the phase increment I, i.e.

$$P(n+1) = Mod_N \left[P(n) + I \right].$$

(5.17)

The look-up table is also called phase-to-amplitude converter. It contains one period of a sinusoidal waveform. The phase $P(n)$ is fed to the look-up table as an address input, under which the amplitude value appears at the output. Multiplied with the factor A_{NBN}, the amplitude values finally form the narrow band noise nbn with its fundamental oscillation. The center frequency f_m depends on the clock frequency f_a, on the number of stored amplitude values N, as well as the on the phase increment I_m, i.e.

$$f_m(I_m) = \frac{f_a}{N} \cdot I_m.$$

(5.18)

The variation of the center frequency can be realized by incrementing or decrementing the phase increment I, i.e.

$$I = \begin{cases} I_m + \Delta I \\ I_m - \Delta I \end{cases}.$$

(5.19)

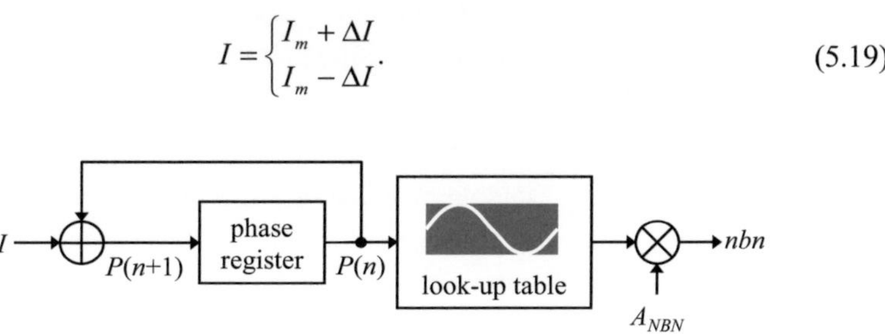

Fig. 5.16 Phase accumulation unit.

When I reaches its lower limit I_{max-}, it will be incremented by ΔI with each clock. Otherwise, when it reaches the upper limit I_{max+}, it is decremented by ΔI with each clock. The resulting instantaneous frequency $f(I)$ is

$$f(I) = \frac{f_a}{N} \cdot I.$$ (5.20)

As a result, the bandwidth can be calculated by

$$\Delta f = \frac{2f_a}{N} \cdot \left(I_{max+} - I_{max-} \right).$$ (5.21)

5.5 Swept-Frequency Noise (SFN)

In addition to the typical narrowband interferers, a class of interferers with time-varying frequencies has been observed both in the in-door and the access domain of NB-PLC. Due to its typical feature in the frequency domain, it is called swept-frequency noise (SFN) in the following sections.

5.5.1 Typical SFN Waveforms and Their STFT

Fig. 5.17 through show some examples of measured SFN waveforms. The noise waveform shown in Fig. 5.17 (a) is filtered by a bandpass FIR filter with a passband between 22 and 35 kHz. The scaled mains voltage is shown in the same plot to illustrate the synchronization of the noise envelope with the mains frequency. Obviously, the noise envelope reaches its maximum at the same time when the mains voltage reaches its absolute peak value. The spectrogram of the STFT in Fig. 5.17 (b) shows two intersecting periodic traces. The first trace starts with 22 kHz at about 2 ms and increases to 32 kHz linearly within 5 ms. The second trace decreases from 30 to 22 kHz in the same time interval.

Fig. 5.18 gives another example obtained in a transformer station. Again, two periodic frequency traces can be observed. Similar to the traces in Fig. 5.17, the frequencies here also change linearly and periodically. Nevertheless, the frequencies sweep between 185 and 210 kHz. There is no overlapping area, i.e. these two traces appear one after the other. In addition, the noise level doesn't show a significant fluctuation over time.

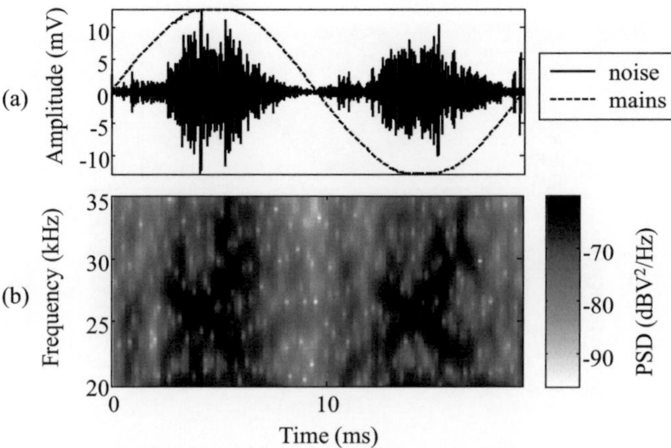

Fig. 5.17 Measured periodic noise with rising and falling swept frequencies at the same time (a) waveform in the time domain (b) spectrogram of STFT

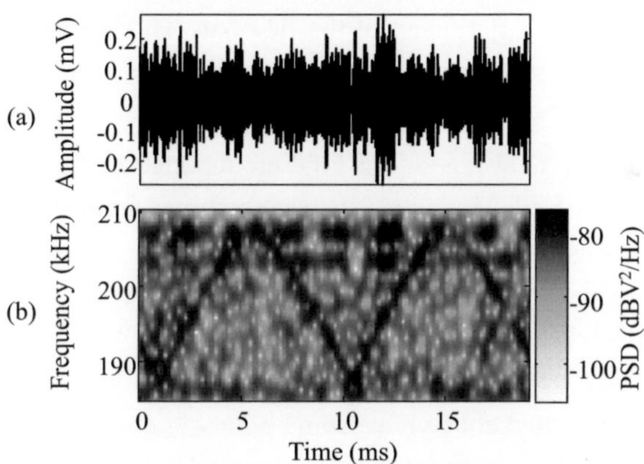

Fig. 5.18 Measured periodic noise with sequential rising and falling swept frequencies (a) waveform in time domain, (b) spectrogram of STFT.

Fig. 5.19 shows a third example. The noise level is much higher than in the first two examples. The waveform looks like a damped oscillation with duration of approx. 2 ms. The spectrogram of the STFT illustrates a complicated pattern. Obviously, the frequency doesn't change linearly over time. Instead, two convex-shaped traces can be observed. The longer trace starts with 140 kHz and decreas-

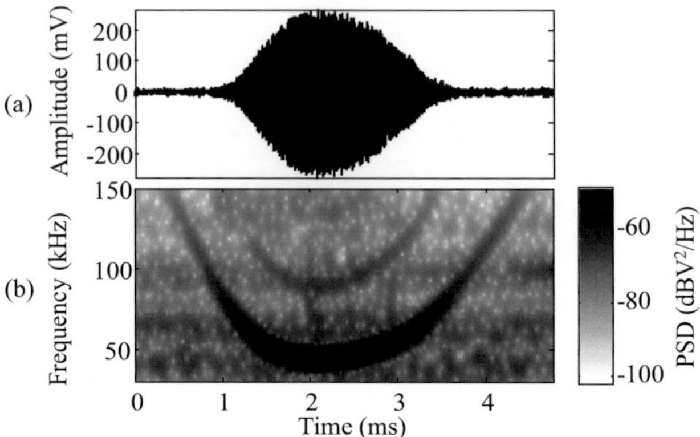

Fig. 5.19 Measured noise with rising and falling swept frequencies at a transformer substation (a) waveform in time domain, (b) STFT.

es to 45 kHz. At the same time the PSD increases to a maximum at 45 kHz. In the second part the frequency increases until it reaches 140 kHz again. The second trace seems to be a harmonic of the first one. This waveform appears as a single noise event in our measurement. However, similar patterns can also occur periodically with a period of 10 ms.

Several periodic noise events of this kind are reported in [LAR11a]. An example is shown in Fig. 5.20. The noise has been recorded in the direct vicinity of a fluorescent lamp after this lamp had been turned on. The filtered noise waveform reaches almost 2 V. The individual oscillations have a quite similar envelope as the one shown in Fig. 5.19 (a). The spectral patterns between 30 and 140 kHz in the spectrogram also match the convex shape shown in Fig. 5.19 (b) quite well.

All examples presented above have some points in common. In the frequency domain, their instantaneous frequencies have small bandwidths but change with time, either linearly as shown in Fig. 5.17 and Fig. 5.18, or nonlinearly such as the pattern in Fig. 5.19 and Fig. 5.20. The sweep bandwidth can range from ten to hundreds of kHz. In the time domain, the envelopes can be periodic and synchronous with the mains frequency, or be relatively constant. They can even be aperiodic and appear as individual lobes with high levels.

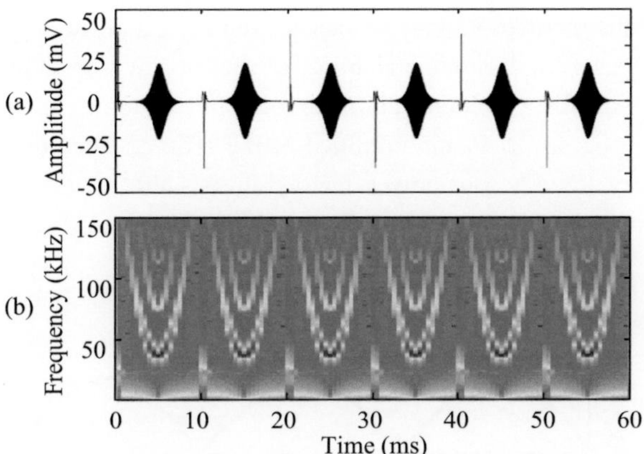

Fig. 5.20 Periodically damped oscillation reported in [LAR11a] (a) periodic waveform in time domain, (b) spectrogram of STFT. Both plots share the same abscissa.

5.5.2 Origin of SFN

One main source of this noise class are active power factor correction (PFC) circuits in power supply units of various end-user appliances, such as fluorescent lamps and PCs. Fig. 5.21 shows a simplified circuit of a switch mode power supply (SMPS) with an inserted active PFC module. In a SMPS without a PFC circuit, the input capacitor C is placed directly behind the rectifier diodes D_1

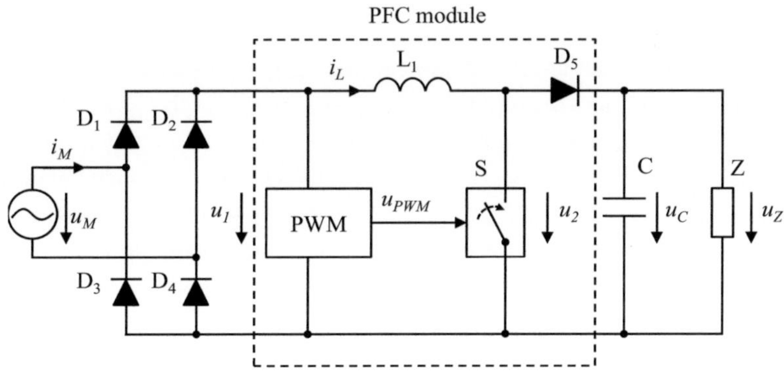

Fig. 5.21 Simplified circuit of a PFC boost pre-regulator [FAI04].

137

through D_4. The current i_M is drawn from the mains to charge C. C will only be charged when the rectified voltage u_1 exceeds the voltage u_C across C. As a result, i_M exhibits large spikes during the charging of the capacitor as shown in Fig. 5.22. The resulting waveform contains a large amount of harmonics in its spectrum. At the same time, the power factor - defined as the ratio of true power to apparent power - is very low. A low power factor burdens the power supply utilities. Moreover, harmonic distortions degrade the power quality and cause EMC problems.

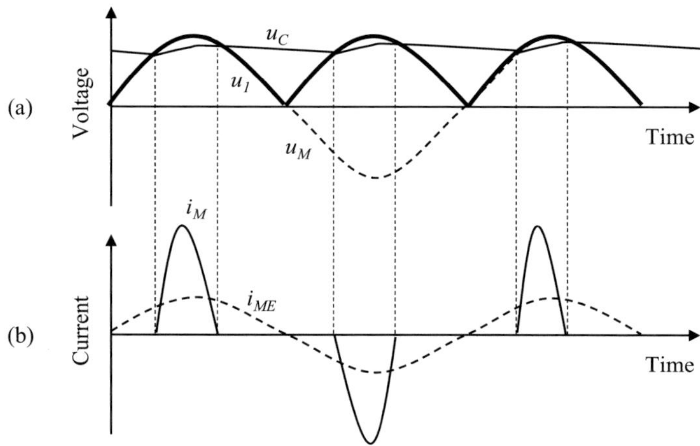

Fig. 5.22 Voltage and current waveforms in a SMPS without a PFC module [TIE09] (a) voltage waveforms, (b) current waveforms; i_{ME} is the expected current waveform for a unity power factor.

The active PFC module is used to keep the current in phase with the mains voltage and to minimize the input current distortion so that the power factor can be raised [FAI04]. As shown in Fig. 5.21, the PFC circuit is actually a boost converter which is mainly composed of an inductor (L), a pulse-width modulator (PWM) and a power MOSFET as a switch (S). The boost converter shall be able to provide a higher voltage u_2 at its output than the peak value of u_1 at its input. Simultaneously, i_L must be well controlled so that i_M is proportional to u_M at any given instant. Implementation details of the control unit can be found in [FAI04] and [TIE09].

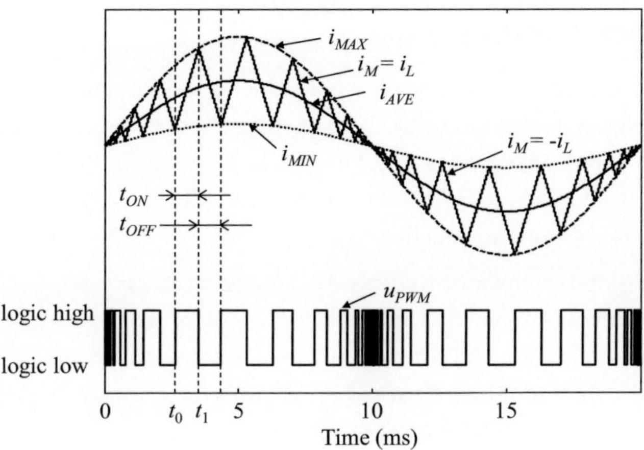

Fig. 5.23 Waveforms of current and PWM control signals generated by a PFC circuit in continuous mode of operation.

Fig. 5.23 shows waveforms of i_M, i_L, and the PWM signal u_{PWM} when the PFC operates in continuous mode. Suppose the inductor L is uncharged initially. When the switch closes at t_0, u_{PWM} becomes logic high.

The inductor current i_L increases linearly according to

$$i_L(t) = i_{MIN}(t_0) + \frac{1}{L} \cdot u_L \cdot (t - t_0), \quad t_0 < t \leq t_{ON} + t_0, \tag{5.22}$$

where u_L is the voltage across L and can be approximated as a constant within t_{ON}, i.e.

$$u_L = u_1, \quad t_0 < t \leq t_{ON} + t_0. \tag{5.23}$$

The switch opens (low level of u_{PWM} within time window t_{OFF}) as soon as $i_L(t)$ reaches $i_{MAX}(t)$ at t_1 and the inductor starts to discharge. Then we have

$$i_L(t) = i_{MAX}(t_1) + \frac{1}{L} \cdot u_L \cdot (t - t_1) \quad t_1 < t \leq t_{OFF} + t_1, \tag{5.24}$$

where u_L is now the voltage difference between u_1 and u_C, i.e.

139

$$u_L = u_1 - u_C, \quad t_1 < t \le t_{OFF} + t_1. \tag{5.25}$$

Due to the nature of a boost converter, u_C shall be greater than u_1. Therefore, u_L is now taking on a negative value, while i_L maintains its direction and decreases linearly. As soon as i_L drops to i_{MIN}, the switch closes again. In this way, the current drawn from mains i_M is kept within the area defined by i_{MAX} and i_{MIN}. Its average value i_{AVE} follows u_m and thus is in phase with the mains voltage.

Depending on the value of i_{MIN}, the PFC module can operate in either discontinuous or continuous mode. In the first mode, i_{MIN} is zero over the entire mains cycle. i_L can reach zero, and the current waveform swings between 0 and i_{MAX}. If i_{MIN} is greater than zero and is synchronized with i_{MAX}, as shown in Fig. 5.23, i_L can never reach zero during the switching cycle.

The HF components in i_M can be converted to voltages by any connected impedance. Although many SMPSs have EMI filters inserted between their rectifier-bridges and the power plugs, these filters are usually not very effective to reduce differential mode noise within the frequency range up to 150 kHz. Therefore, most spectral components of the noise can still appear in the mains voltage and be coupled into NB-PLC systems.

Fig. 5.24 (a) shows a bandpass filtered waveform of i_M. The spectrogram of the STFT is shown in (b). Similarities can be observed in both the noise waveform

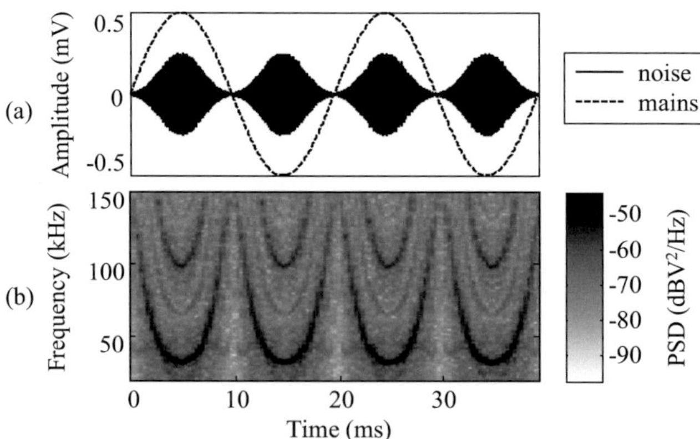

Fig. 5.24 HF components of synthesized i_M obtained by applying a bandpass filter. Passband: 20-150 kHz, (a) filtered waveform, (b) spectrogram of the STFT.

and the spectral characteristics between the synthesized noise and the measured noise events shown in Fig. 5.19 and Fig. 5.20.

More and more active PFC modules are being applied to reduce the harmonics generated by end-user devices and to comply with international standards, such as IEC 61000-3-2. Furthermore, active PFC circuits are the most favorable solutions to limit harmonics for lighting equipment with HF-ballast [LAR11A]. Therefore, the influence of SFN on the performance of NB-PLC systems will become larger. It is important to improve the robustness of communication systems against SFN. Moreover, it is necessary to add this noise class to the current noise scenario and to emulate it as accurate as possible.

5.5.3 Emulating SFN using Chirps

Due to the distinct spectral features of the SFN, it is reasonable to treat SFN in the frequency domain. The SFN can be modeled by a superposition of multiple chirp functions, such as

$$y_{chirp} = \sum_{n=1}^{N} m_n(t) \cdot \sin\left[2\pi \int_0^t f_n(\tau)d\tau \right], \tag{5.26}$$

where N is the number of fundamental chirp waveforms, $m_n(t)$ and $f_n(t)$ are the envelope and the instantaneous frequency of the nth chirp respectively. For the periodical noise with superimposed linear chirps and time varying envelopes, such as those in Fig. 5.17, the instantaneous frequency is obtained by

$$f_n(\tau) = f_0 + \frac{f_1 - f_0}{\Delta\tau} \cdot \tau, \tag{5.27}$$

where f_0 and f_1 are the start and stop frequencies respectively, and $\Delta\tau$ is the duration within which the frequency changes from f_0 to f_1. $m_n(t)$ can be approximated by the scaled waveform of the mains voltage. The emulated noise together with colored background noise is shown in Fig. 5.25. The noise in Fig. 5.18 is composed of two cascaded basic linear chirps. Both of them can be modeled using (5.26) and (5.27). Fig. 5.26 shows the emulated counterpart.

The chirp in Fig. 5.19 has an instantaneous frequency which changes nonlinearly following a concave shape

$$f_n(\tau) = f_1 \cdot \frac{(\tau - \tau_0)^{1.8}}{\tau_0^{1.8}} + f_0, \tag{5.28}$$

where τ_0 is the time corresponding to the minimum frequency f_0. f_1 is the maximum frequency. The harmonic frequencies can also be modeled by (5.28) conveniently. The envelope can be approximated by a Chebyshev window function

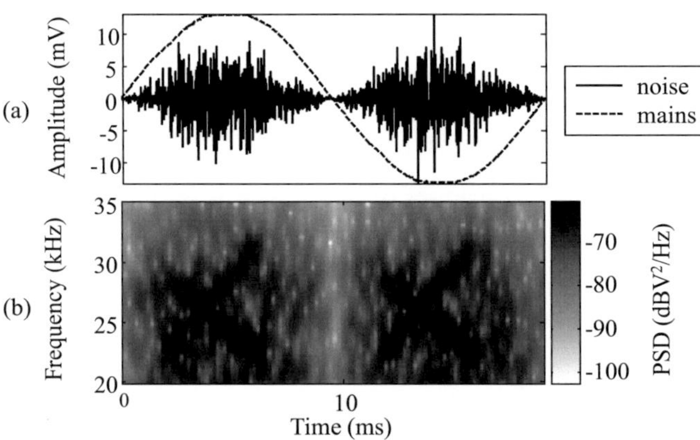

Fig. 5.25 Synthesized periodic noise with rising and falling swept frequencies. (a) waveform in the time domain, (b) STFT of the waveform.

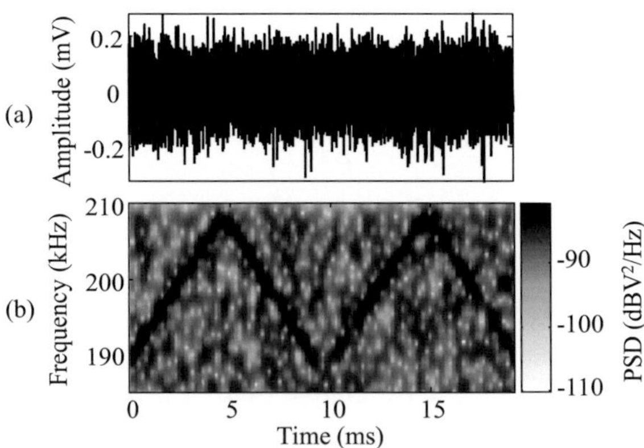

Fig. 5.26 Synthesized periodic noise with rising and falling swept frequencies. (a) waveform in the time domain, (b) STFT of the waveform.

whose Fourier transform sidelobe magnitude is 120 dB below the main lobe magnitude. The synthesized noise waveform is shown in Fig. 5.27.

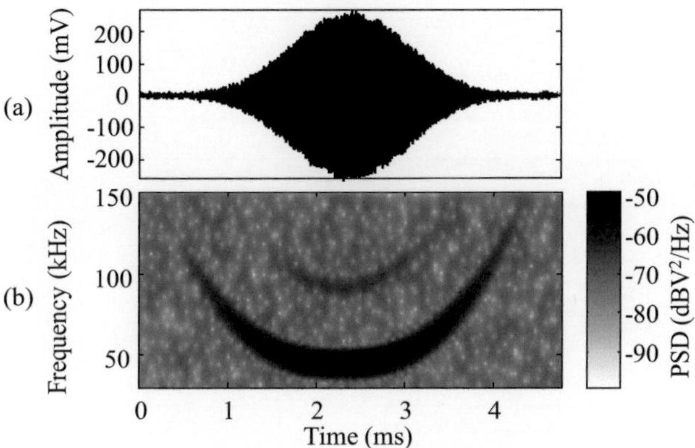

Fig. 5.27 Synthesized periodic noise with rising and falling swept frequencies. (a) waveform in the time domain, (b) STFT of the waveform.

5.5.4 Emulating SFN Using a PFC Control Algorithm

Another idea is to synthesize the current $i_M(t)$ using a PFC control algorithm and to filter it using a bandpass filter. Fig. 5.28 shows a flow chart for the control algorithm. The constants A, B and Δi are used to define the SFN. A and B are scaling factors used to define peak and minimum currents. Factor A must be smaller than B. Δi determines the step size for $i_L(t)$ when it increases or decreases. Actually, the step size for the current increase can also be different from the one for current decrease. Here the same value is used for simplicity.

The algorithm begins with $i = 1$. A small value $|m(1)/B|$ is assigned to $i_M(1)$. A variable, denoted by $f_{up}(k)$, indicates whether the inductor current $i_L(k)$ is increasing or decreasing . It reflects the state of the switch S in Fig. 5.23. Value 1 indicates that the switch is closed, and the inductor is being charged while $i_L(k)$ increases. On the contrary, value 0 means the switch is open, the inductor is being discharged and $i_L(k)$ decreases. It is assumed that the inductor is not charged and the switch closes at the moment when the control process begins. Therefore, value 1 is assigned to $f_{up}(k)$ initially. The peak and minimum currents, denoted by $\hat{i}_{MAX}(k)$ and $\hat{i}_{MIN}(k)$ here, are obtained by dividing the absolute value of mains

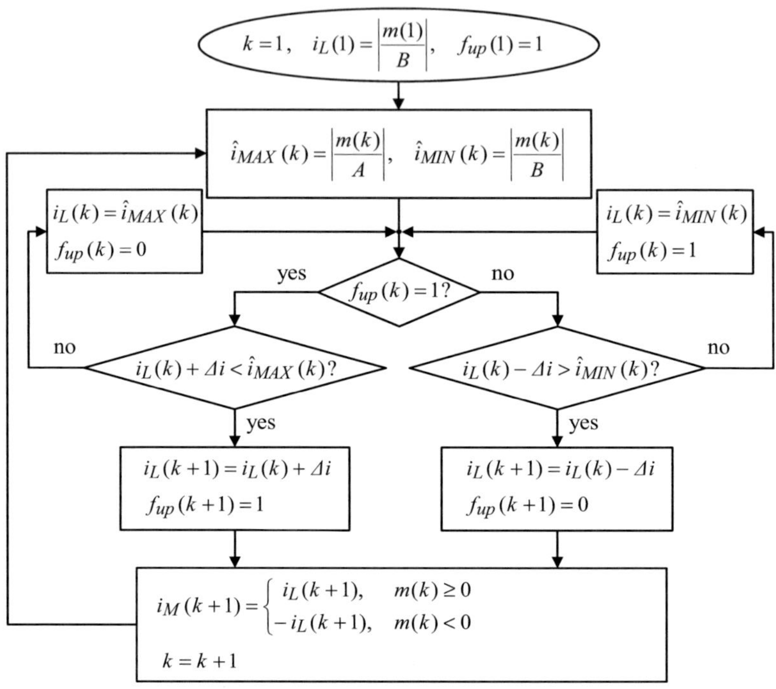

Fig. 5.28 Flow chart of the PFC control algorithm for the generation of $i_M(t)$

voltage $m(k)$ by the factors A and B respectively. The reason why the absolute value instead of the signed mains voltage is used will be explained shortly. Recall the two current waveforms $i_{MAX}(k)$ and $i_{MIN}(k)$ in Fig. 5.23. They were directly proportional to the mains voltage. Therefore they were positive in the first half of a mains cycle and negative in the second half cycle. The control process for both half cycles is almost the same, except for the part associated with $i_{MAX}(k)$ and $i_{MIN}(k)$. In practice, it is much more convenient to treat the second half cycle in the same manner as the first half at first, and then match the sign of the final result. Therefore, the negative halves of $i_{MAX}(k)$ and $i_{MIN}(k)$ are mirrored, so that the control process remains the same for both half cycles.

After having assigned values to $\hat{i}_{MAX}(k)$ and $\hat{i}_{MIN}(k)$, the state of the switch is checked. If $f_{up}(k)$ has the value 1, the switch is closed, otherwise it is open. In the case of a closed switch, provided that an increase of i_L by Δi will not lead to an excess of $\hat{i}_{MAX}(k)$, then i_L will add Δi to itself and the switch remains closed. Otherwise, if $i_L(k) + \Delta i > \hat{i}_{MAX}(k)$, then i_L will be clipped to $\hat{i}_{MAX}(k)$, and the switch will be opened. In the case of an open switch, i_L will be reduced by Δi. If the new value drops below $\hat{i}_{MIN}(k)$, then i_L will be clipped to $\hat{i}_{MIN}(k)$, and the switch will be closed. Otherwise, if the new value is larger than $\hat{i}_{MIN}(k)$, then the new value will be assigned to $i_L(k+1)$ and the switch keeps open. As soon as $i_L(k+1)$ is determined, the sign of $m(k)$ will be checked. For a positive value, the expected current $i_M(k+1)$ is the same as $i_L(k+1)$. Otherwise, $i_M(k+1)$ has an opposite direction as $i_L(k+1)$. An example of the resulting SFN is shown in Fig. 5.24.

5.6 Impulsive Noise

Impulsive noise is usually classified into three categories: *periodic impulsive noise asynchronous to the mains frequency* mainly caused by switched power supplies, p*eriodic impulsive noise synchronous to the mains frequency* caused by the switching of rectifier diodes of power supplies and *asynchronous impulsive noise* originating from transients caused by switching events. Setting aside the time behavior, such as inter-arrival time, individual impulses have some features in common.

In the time domain, the impulses are peaks with short duration and high amplitudes. In contrast to the typical random waveforms of background noise, most impulsive noise types exhibit deterministic appearance patterns, such sharp rising edges followed by damped oscillations, low-level oscillations terminated by sharp endings [ZIM02], or impulse chains, either equally spaced or not - see e.g. [COR10].

In the frequency domain, impulsive noise can be distinguished from background noise by a raised wideband PSD. Most measured impulses exceed the background noise power spectral density for at least 10-15 dB within most portions of the frequency range. These common features apply not only for broadband PLC but also for NB-PLC channels. The broadband portion of the noise spectrum is mainly caused by the sharp rising edges of the impulses - see e.g. [ZIM02].

Fig. 5.29 shows a waveform dominated by impulsive noise together with background noise and the corresponding instantaneous powers in plot (a) and plot (b) respectively. Plot (c) is a detailed view of the part between 16 and 17.5 ms. The

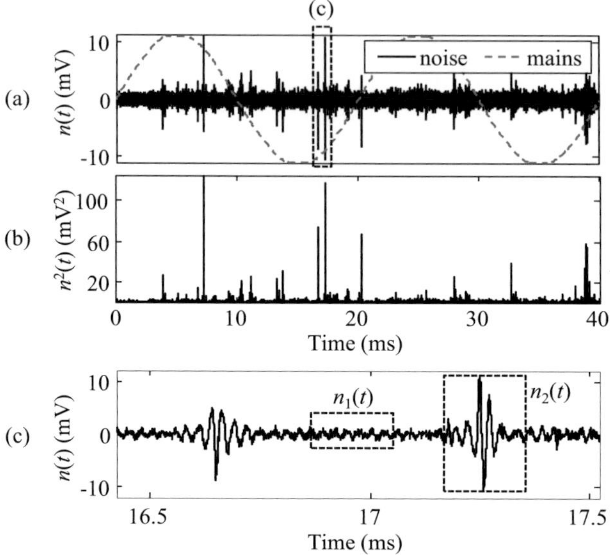

Fig. 5.29 Measured noise in the time domain, (a) noise waveform n(t), (b) $n^2(t)$, (c) detailed view of n(t), corresponding to part (c) in plot (a).

difference of the levels between the background noise $n_1(t)$ and the impulse $n_2(t)$ is straightforward in plot (c). The peak value of $n^2(t)$ exceeds the average level of $n_1(t)$ by several orders of magnitude. Fig. 5.30 shows the PSDs and the phase vectors for $n_1(t)$ and $n_2(t)$ in plot (a) and plot (b) respectively. Obviously, the PSD of $n_2(t)$ exceeds the PSD of $n_1(t)$ by more than 10 dB at most frequencies. Unlike the PSD, the phase vectors of the impulsive noise are rarely investigated in literature. Actually, impulsive noise shows quite different phase vectors than the colored background noise. As shown in plot (b), the phase vector of $n_1(t)$ exhibits a small random profile near zero degrees, while the phase response of $n_2(t)$ increases or decreases with frequency with recognizable patterns.

5.6.1 Detecting and Removing Impulsive Noise

The goal of this section is to show a way to detect individual impulses - regardless of the noise type - and remove them from the original noise. The detection is made by observing the aforementioned time-frequency characteristics and by executing the following steps. Suppose the noise spectrum is expressed by

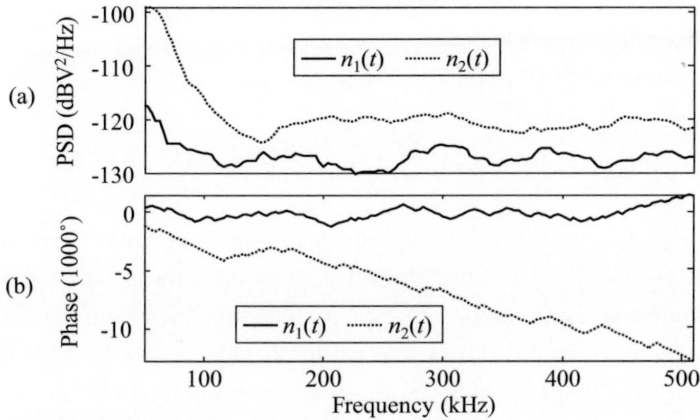

Fig. 5.30 PSD and phase of noise segments $n_1(t)$ and $n_2(t)$, (a) PSD, dashed lines are original magnitudes, solid lines are curve-fitted using a least square procedure; (b) unwrapped phase vector given in degrees.

$$A(m,n) = |A(m,n)| \cdot \exp\left[j \cdot \varphi(m,n) \right], \tag{5.29}$$

where m is segment index of the waveform in the time domain, and n is the index of the frequency bin. $|A(m,n)|$ and $\varphi(m,n)$ are the magnitude and phase vectors at the nth frequency bin in the mth time interval.

Suppose the length of the waveform segment is denoted by T_{seg}, then the normalized noise energy is

$$P(m) = \frac{\Delta f \cdot \sum_{n=0}^{N-1} |A(m,n)|^2 \cdot T_{seg}}{\max\left(\Delta f \cdot T_{seg} \cdot \sum_{n=0}^{N-1} |A(m,n)|^2 \right)}, \tag{5.30}$$

where N is the total number of frequency bins. To locate the high energy peaks, $P(m)$ is compared with a threshold P_{th}. The noise segments where the condition

$$P\left(m_{peak} \right) > P_{th} \tag{5.31}$$

is fulfilled, are considered to be dominated by impulsive noise events. $A(m_{peak}, n)$ contains spectral components of both background and impulsive noise. The spectrum of the noise segment in the time interval m_{peak} is

$$A\left(m_{peak}, n\right) = \left|A\left(m_{peak}, n\right)\right| \cdot \exp\left[j \cdot \varphi\left(m_{peak}, n\right)\right].$$

(5.32)

As mentioned in section 5.2, most impulsive noise originates from sharp high-energy pulses. The PSD and the phase vector of each measured impulse reflect the magnitude and phase responses of the propagation trace from the noise source to the measurement spot. Lessons learned in chapter 3 show that the magnitude responses of the NB-PLC channels are usually smooth and change slowly with frequency. This feature allows application of curve fitting approaches in estimating the PSDs of impulsive noise. Curve-fitting methods based on polynomial models or smoothing methods can be applied to the power spectrum $|A(m_{peak}, n)|^2$ along the frequency axis as follows:

$$\left|\hat{A}(m_{peak}, n)\right|^2 = Poly\left(\left|A(m_{peak}, n)\right|^2, N_P\right),$$

(5.33)

where N_P denotes the order of the polynomial. Applying the same curve fitting method to the PSDs of the neighboring background noise segment leads to

$$\left|\hat{A}_{BGN}\left(m_{peak} - \Delta m, n\right)\right|^2 = Poly\left(\left|A_{BGN}(m_{peak} - \Delta m, n)\right|^2, N_P\right),$$

(5.34)

where Δm denotes the time interval between a noise impulse and its neighboring background noise. The difference between $|\hat{A}(m_{peak}, n)|^2$ and $|\hat{A}_{BGN}(m_{peak} - \Delta m, n)|^2$ - denoted by ΔP - is obtained by

$$\Delta P\left(m_{peak}, n\right) = \left|\hat{A}\left(m_{peak}, n\right)\right|^2 - \left|\hat{A}_{BGN}\left(m_{peak} - \Delta m, n\right)\right|^2.$$

(5.35)

The PSD of the impulsive noise is approximated by

$$\left|\hat{A}_{IMP}\left(m_{peak}, n\right)\right|^2 \approx \begin{cases} \Delta P\left(m_{peak}, n\right), & \Delta P\left(m_{peak}, n\right) > P_{th} \\ 0, & \Delta P\left(m_{peak}, n\right) \le P_{th} \end{cases}.$$

(5.36)

In this approximation, $|\hat{A}_{BGN}(m_{peak}-\Delta m, n)|^2$ is used to estimate $|\hat{A}_{BGN}(m_{peak}, n)|^2$. At frequencies where ΔP exceeds a threshold P_{th}, it can be assumed that each $|\hat{A}(m_{peak}, n)|^2$ is a numerical sum of $|\hat{A}_{IMP}(m_{peak}, n)|^2$ and $|\hat{A}_{BGN}(m_{peak}, n)|^2$. Therefore $\Delta P(m_{peak}, n)$ is assigned to $|\hat{A}_{IMP}(m_{peak}, n)|^2$. At frequencies where ΔP is smaller than P_{th}, $|\hat{A}(m_{peak}, n)|^2$ are considered to be dominated by the spectral components of the background noise. In this case $|\hat{A}_{IMP}(m_{peak}, n)|^2$ takes on the value 0. Fig. 5.31 (b) illustrates the PSD difference and the spectrum of the impulsive noise obtained according to (5.36).

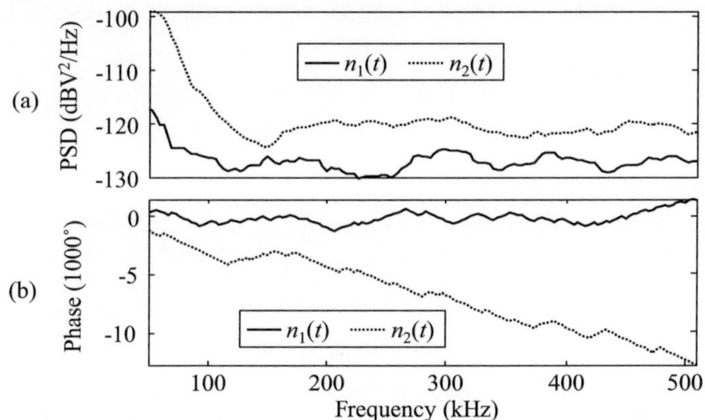

Fig. 5.31 Estimation of power spectrum for impulsive noise. (a) PSD of a noise segment containing both impulsive and background noise, and PSD of neighboring background noise segments, (b) $\Delta P(m_{peak}, n)$ and resulting PSD of the impulsive noise. Most parts of both curves overlap.

The waveform of the impulse in the time interval m_{peak} is obtained by performing an IFFT of the $\hat{A}_{IMP}(m_{peak}, n)$, i.e.

$$y\left(m_{peak}, t\right) = \sum_{n=0}^{N-1} \left|\hat{A}_{IMP}\left(m_{peak}, n\right)\right| \cdot \exp\left[j \cdot \varphi\left(m_{peak}, n\right)\right] \cdot \exp\left[j \cdot \frac{2\pi n t}{N}\right]. \qquad (5.37)$$

Theoretically, the influence of the background noise in the phase term should also be eliminated to obtain an error-free reconstruction of the impulsive noise. However, as the impulsive noise has higher magnitudes at most frequencies, the phase errors caused by the background noise are relatively small. Therefore, the phase term $\varphi(m_{peak}, n)$ is applied in (5.37) for simplification. Fig. 5.32 shows an example of the extracted noise pattern.

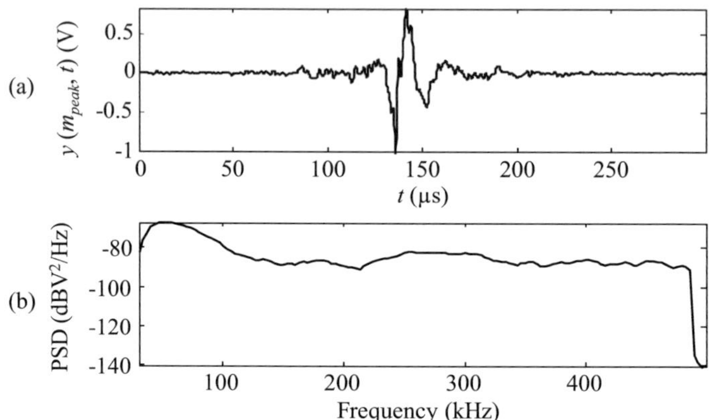

Fig. 5.32 Pattern of impulsive noise. (a) waveform, (b) spectrum.

All the impulsive waveforms can be obtained by applying this algorithm to the selected intervals m_{peak}. The reconstructed patterns can be subtracted from the original noise waveform. This way the remaining waveform represents pure background noise. Fig. 5.33 (a) and (b) show waveforms before and after the re-

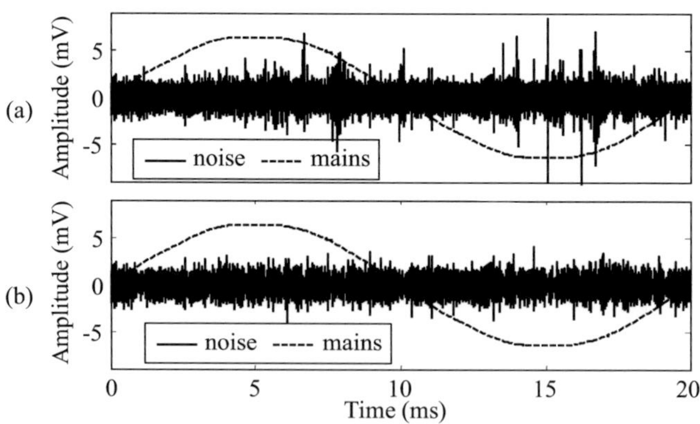

Fig. 5.33 Comparison of noise waveforms before and after removal of impulsive noise. (a) original noise waveform, (b) noise waveform after removal of impulses.

moval of the impulsive noise portions respectively. It can be seen that all significant impulses are removed. At the same time, the variance of the background noise is not affected.

5.6.2 Emulating Time-Domain Waveforms of Impulsive Noise

In the case where the time-domain noise waveform plays a less important role as the noise PSD, it can be assumed that the impulsive noise has a random waveform, and the appearance of such an impulse only raises the spectral level of the background noise shortly. This kind of impulsive noise can be generated by switching the background noise on and off according to a control pulse sequence. As shown in Fig. 5.34, colored background noise serves as noise source. The scaling unit determines the maximum amplitude of the noise. A switch is inserted between the noise source and the scaling unit. The switching sequence generator launches a sequence of '1' or '0'. '1' switches the noise source on to the scaling unit. The width of the pulse is determined by the duration of a '1' level. If the '1' level repeats within constant intervals, periodic impulsive noise is produced. Otherwise we get aperiodic impulsive noise. The former can be defined as synchronous with the mains frequency, if it repeats e.g. every 10 or 20 ms. All other rates correspond to periodic impulsive noise asynchronous with the mains frequency. The emulation of the time behavior will be introduced later in the next section.

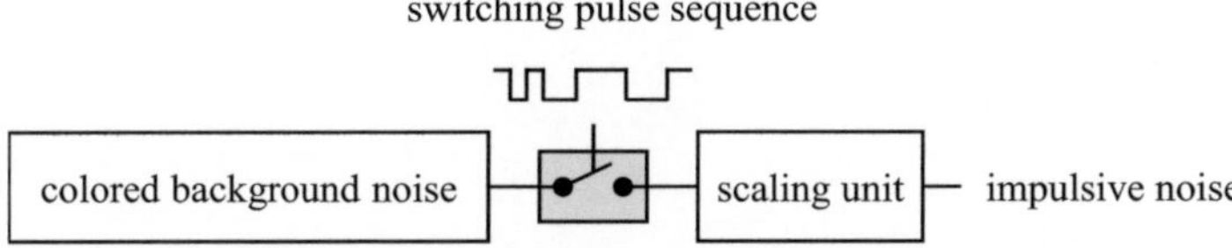

Fig. 5.34 Generation of impulsive noise with random waveforms.

In case where an oscillation with exponential decay is expected, the impulsive waveform can be controlled by using three parameters: the pulse width, the maximal amplitude and the frequency of the oscillation. The noise can be modeled by modulating a sinusoidal waveform according to

$$y_0 = \sin\left(\omega_0 \cdot t\right), \tag{5.38}$$

with an exponentially attenuated envelope given by

$$x_0 = A_0 \cdot e^{-a_0 \cdot t}, \tag{5.39}$$

where $\omega_0 = 2\pi f_0$ is the circular frequency of the sinusoidal signal. A_0 is the initial amplitude, and a_0 is the time constant of the exponential decay. The modulated signal can be expressed as

$$z_0 = x_0 \cdot y_0 = A_0 \cdot e^{-a_0 \cdot t} \cdot \sin(\omega_0 \cdot t). \tag{5.40}$$

The implementation of the noise is shown in Fig. 5.35. A segment of an exponential function with 1000 samples as well as a period of a sinusoidal signal with the same number of samples is created in MATLAB first and stored as a look-up table within the FPGA hardware. The step width of the generated addresses can be configured with the parameters Δm and Δn. x_1 and y_1 are outputs of both look-up tables. They are multiplied with each other, and the product sequence forms the modulated exponentially decaying oscillation z.

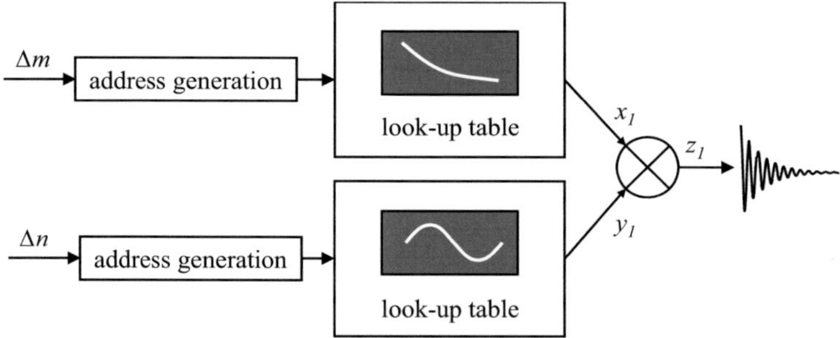

Fig. 5.35 Emulation of impulsive noise with exponentially decaying waveform.

Combining both into the expressions for both signals, we get

$$x_1 = A_0 \cdot e^{-a_0 \cdot (\Delta m \cdot T_s)} = A_0 \cdot e^{-a_1 \cdot T_s}, \tag{5.41}$$

and

$$y_1 = B_1 \cdot \sin(\omega_0 \cdot \Delta n \cdot T_s) = B_1 \cdot \sin(\omega_1 \cdot T_s), \tag{5.42}$$

where T_s is the period of the clock, with which samples are read out of the look-up tables. Changing the step widths of both addresses provides a variation of the sampling period. This can also be interpreted as an alteration of the time constant of the exponential decay and the frequency of the oscillation respectively. This way, both signals can be configured conveniently.

5.6.3 Emulating Periodic Time Behavior

Three parameters are of importance for periodic impulsive noise: the repetition rate r_{PIN}, the pulse width d_{PIN}, and the amplitude A_{PIN}. As mentioned above, A_{PIN} is manipulated by the scaling unit, and only the first two parameters are determined by the switching pulse sequence. The periodic switching sequence is generated by a modulo-N counter, which counts from 0 to N-1, as shown in Fig. 5.36.

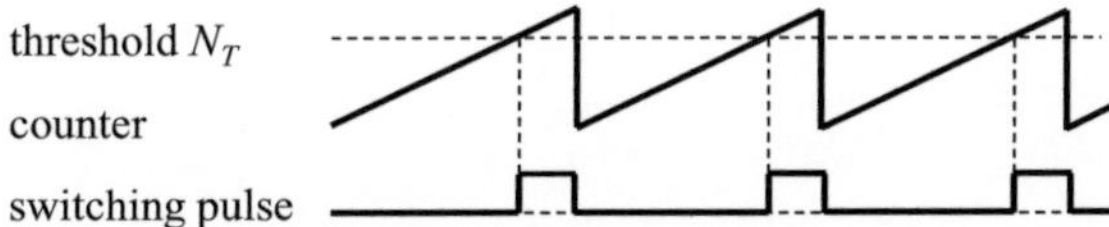

Fig. 5.36 Generation of a switching sequence.

The counter will be reset to 0 by the next clock impulse after it has reached N-1. The counter value is compared with a threshold N_T. As soon as the count value is greater than N_T, the switching signal is asserted. Since the threshold is constant, a periodical sequence of pulses is generated. The repetition rate r_{PIN} is

$$r_{PIN} = \frac{f_0}{N},$$

(5.43)

where f_0 is the clock frequency. The pulse width can be calculated by

$$d_{PIN} = \frac{N - N_T}{f_a}.$$

(5.44)

5.6.4 Emulating Time Behavior for Aperiodic Noise

The time behavior of aperiodic impulsive noise (APIN) features stochastic characteristics. In [ZIM02] the parameters impulse amplitude A_i, impulse width t_W, impulse distance t_D, and arrival time t_{ARR} have been introduced to describe the statistical characteristics in the form

$$y_{APIN}(t) = \sum_{i=1}^{N} A_i \cdot imp\left[\frac{t - t\left(t_{ARR,i} + t_{W,i}/2\right)}{t_{W,i}} \right] \tag{5.45}$$

where y_{APIN} denotes the impulse envelope in the time domain. $imp(t)$ is a rectangular function of time for an individual envelope. N denotes the total number of impulses and i is the index of each impulse. It can be seen in Fig. 5.37 that the parameters fulfill

$$t_{ARR,i+1} = t_{ARR,i} + t_{W,i} + t_{D,i}. \tag{5.46}$$

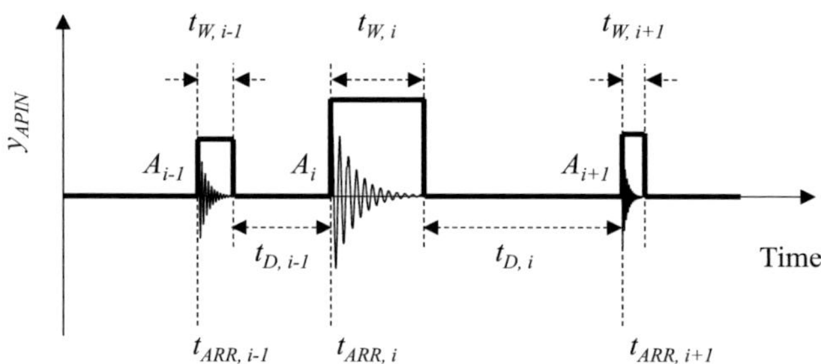

Fig. 5.37 Envelope of aperiodic impulsive noise in the time-domain. Units for abscissa and ordinate are omitted for simplicity.

The impulse rate r_{APIN} is defined as the ratio of the number of impulses N_{APIN} to the corresponding observation window T_{WIN}, i.e.

$$r_{APIN} = \frac{N_{APIN}}{T_{WIN}}. \tag{5.47}$$

The disturbance ratio is defined by the ratio of the total impulse width to the observation window:

$$r_d = \frac{\sum_{i=1}^{N_{APIN}} t_{W,i}}{T_{WIN}}.$$

(5.48)

The parameters A_i, $t_{W,i}$ and $t_{D,i}$ are random variables. Measurements at real-world channels have shown that the distributions of both $t_{W,i}$ and $t_{D,i}$ can be approximated by a superposition of multiple exponential distributions. A partitioned Markov chain with multiple states was proposed for proper modeling of the time behavior in [ZIM02]. The partitioned Markov chain approach is based on basic Markov chains. A Markov chain is basically a stochastic process with the following so-called Markov property: For a given present state, the probabilities of future states are always independent of the past states. The description of the present state fully captures all the information that could influence the future evolution of the Markov process. Future states will be reached through a probabilistic process instead of a deterministic one. At each step, the state may remain the same, or may change from the current state to another one, according to a probability distribution. The changes of states are called transitions, and the probabilities associated with all possible state-changes are called transition probabilities. The statistical properties of a Markov chain are completely described by its transition probability matrix

$$P = \begin{bmatrix} p_{1,1} & p_{1,2} & \cdots & & p_{1,n} \\ p_{2,1} & p_{2,2} & & \ddots & \vdots \\ \vdots & \ddots & \ddots & & p_{n-1,n} \\ p_{n,1} & \cdots & & p_{n,n-1} & p_{n,n} \end{bmatrix},$$

(5.49)

where $p_{i,j}$ is a positive number smaller than 1. It expresses the transition probability from state i to state j, with $i, j = 1, 2, \ldots, n$. The sum of the elements for each row satisfies

$$\sum_{j=1}^{n} p_{i,j} = 1 \qquad i = 1, 2, \ldots, n.$$

(5.50)

In the partitioned Markov chain, the states are divided in two groups, as shown in Fig. 5.38. Group A contains N_D noise-free states, and group B is a collection of N_W states in which impulsive noises appear.

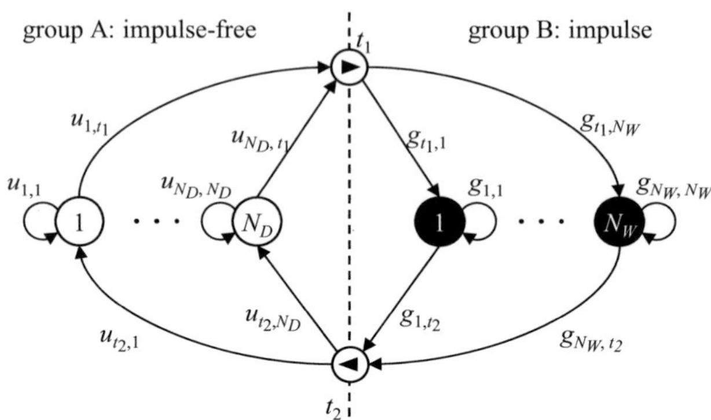

Fig. 5.38 Partitioned Markov chain to model the time behavior of aperiodic impulsive noise.

The transition probabilities in group A and B are denoted by $u_{i,j}$ and $g_{i,j}$ respectively. Besides the two groups, two additional transition states - denoted by t_1 and t_2 - are introduced to describe the crossing from one group to the other. Actually, t_1 and t_2 are not parts of the P-matrix like $u_{i,j}$ and $g_{i,j}$. Instead, they are used to simplify the expression of the state switch from one group to the other. In principle, a transition between states within the same group is not allowed. Any transition from a state in group A to a state in group B must be bridged by t_1. Similarly, any transition from a state in group B to a state in group A must be bridged by t_2. $u_{i,t1}$ denotes the probability of a transition from an impulse-free state i to the transition state t_1.

$$u_{i,t_1} = 1 - u_{i,i}. \tag{5.51}$$

$g_{i,t2}$ denotes the probability of the transition from an impulse state i to the transition state t_2.

$$g_{i,t_2} = 1 - g_{i,i}. \tag{5.52}$$

The average impulse distance $\bar{t}_D$ is obtained by

$$\bar{t}_D = t_a \cdot \sum_{k=1}^{\infty} k \cdot u_{i,i}^{k-1} \cdot \left(1 - u_{i,i}\right) = t_a \cdot \frac{1}{1 - u_{i,i}}. \tag{5.53}$$

Similarly, the average impulse width $\bar{t}_W$ is

$$\bar{t}_W = t_a \cdot \sum_{k=1}^{\infty} k \cdot g_{i,i}^{k-1} \cdot \left(1 - g_{i,i}\right) = t_a \cdot \frac{1}{1 - g_{i,i}}. \tag{5.54}$$

The probability that the Markov chain will remain in the impulse-free states in the kth step is

$$F_A(k) = \sum_{i=1}^{u} u_{t_2,i} \cdot u_{i,i}^{k}. \tag{5.55}$$

Similarly, the probability that the Markov chain stays in the impulse states (group B) for the whole duration of k steps is

$$F_B(k) = \sum_{i=1}^{w} g_{t_1,\,i} \cdot g_{i,\,i}^{k}. \tag{5.56}$$

As a result, the transition probability matrix of the portioned Markov chain is modified into

$$P = \begin{bmatrix}
u_{1,1} & \cdots & 0 & u_{1,t_1} \cdot g_{t_1,1} & \cdots & u_{1,t_1} \cdot g_{t_1,N_W} \\
\vdots & \ddots & \vdots & \vdots & \ddots & \vdots \\
0 & \cdots & u_{N_D,N_D} & u_{N_D,t_1} \cdot g_{t_1,1} & \cdots & u_{N_D,t_1} \cdot g_{t_1,N_W} \\
g_{1,t_2} \cdot u_{t_2,1} & \cdots & g_{1,t_2} \cdot u_{t_2,N_D} & g_{1,1} & \cdots & 0 \\
\vdots & \ddots & \vdots & \vdots & \ddots & \vdots \\
g_{N_W,t_2} \cdot u_{t_2,1} & \cdots & g_{N_W,t_2} \cdot u_{t_2,N_D} & 0 & \cdots & g_{N_W,N_W}
\end{bmatrix}. \tag{5.57}$$

5.6.5 Estimation of a P-Matrix Using Measured Distributions

In some situations, the noise scenario is supposed to be emulated for a specific PLC channel. In such a case, it is common sense to investigate the statistic properties of the channel first. As a direct outcome the distributions of t_W and t_D can be obtained. Suppose the measured distribution of t_D has the form

$$F_{t_D}(k) = \sum_{i=1}^{N_D} a_i \cdot b_i^k ,$$

(5.58)

where N_D denotes the number of exponential distributions for the impulse-free states, and the parameters a_i and b_i for each exponential function can be estimated by applying a simplex method as to be found e.g. in [NEL65]. Recalling the probability function (5.55) for the dwelling of the Markov chain in impulse-free states, then $u_{v+1,i}$ and $u_{i,i}$ correspond to a_i and b_i respectively. Similarly, $g_{w+1,i}$ and $g_{i,i}$ can be obtained by estimating the parameters of the distribution function of t_W, according to

$$F_{t_W}(k) = \sum_{i=1}^{N_W} c_i \cdot d_i^k ,$$

(5.59)

where N_W denotes the number of exponential distributions for the impulse states. The parameters c_i and d_i can be estimated by applying the same approach. As a result, the P-matrix can be obtained as

$$P = \begin{bmatrix} b_1 & \cdots & 0 & (1-b_1) \cdot c_1 & \cdots & (1-b_1) \cdot c_{N_W} \\ \vdots & \ddots & \vdots & \vdots & \ddots & \vdots \\ 0 & \cdots & b_{N_D} & (1-b_{N_D}) \cdot c_1 & \cdots & (1-b_{N_D}) \cdot c_{N_W} \\ (1-d_1) \cdot a_1 & \cdots & (1-d_1) \cdot a_{N_D} & d_1 & \cdots & 0 \\ \vdots & \ddots & \vdots & \vdots & \ddots & \vdots \\ (1-d_{N_W}) \cdot a_1 & \cdots & (1-d_{N_W}) \cdot a_{N_D} & 0 & \cdots & d_{N_W} \end{bmatrix} .$$

(5.60)

5.6.6 Estimation of a P-Matrix Using Desired Noise Parameters

In some applications, impulsive noises has to be synthesized according to desired impulse parameters like $\overline{t}_{W,i}$, $\overline{t}_{D,i}$ and the relative disturbance ratio r_D. With the help of these parameters, the matrix elements $u_{i,i}$, $u_{i,i1}$, $u_{i2,i}$, $g_{i,i}$, $g_{i,i2}$ and $g_{i2,i}$ can be determined. According to (5.53) and (5.54), the matrix elements $u_{i,i}$ and $g_{i,i}$ can be obtained by

$$u_{i,i} = 1 - \frac{t_a}{t_{D,i}}, \qquad i = 1,\ldots N_D, \tag{5.61}$$

and

$$g_{i,i} = 1 - \frac{t_a}{t_{W,i}}, \qquad i = 1,\ldots N_W. \tag{5.62}$$

In the next step, $u_{i,t1}$ and g_{i,t_2} can be obtained by applying (5.51) and (5.52) correspondingly. The stationary distribution π in an irreducible aperiodic Markov chain is given by

$$\pi = \left[\pi_{u,1}, \pi_{u,2}, \ldots, \pi_{u,N_D}, \pi_{g,1}, \pi_{g,2}, \ldots, \pi_{g,N_W} \right], \tag{5.63}$$

where $\pi_{u,i}$ and $\pi_{g,i}$ denote the positive stationary probability for the ith state in the impulse-free and the impulse groups respectively. The elements of the stationary distribution satisfy

$$\pi_{u,i} = \sum_{j=1}^{N_D} \pi_{u,j} \cdot u_{j,i} + \sum_{j=1}^{N_W} \pi_{g,j} \cdot g_{j,t_2} \cdot u_{t_2,i}, \tag{5.64}$$

and

$$\sum_{i=1}^{N_D} \pi_{u,i} + \sum_{i=1}^{N_W} \pi_{g,i} = 1. \tag{5.65}$$

The disturbance ratio r_D can be expressed by

$$r_D = \frac{\displaystyle\sum_{i=1}^{N_W} \pi_{g,i}}{\displaystyle\sum_{i=1}^{N_D} \pi_{u,i} + \sum_{i=1}^{N_W} \pi_{g,i}} = \sum_{i=1}^{N_W} \pi_{g,i}. \tag{5.66}$$

Since the average noise intensity is mainly defined by the disturbance ratio, it is of less importance to know how the distribution values are assigned to the individual states within each group. It is sufficient for the noise synthesis to make sure that $\pi_{u,i}$ and $\pi_{g,i}$ fulfill (5.65) and (5.66). After that, they can be used to calculate the probabilities $u_{t2,i}$ and $g_{t1,i}$ for transitions from t_1 to impulse states and from t_2 to impulse-free states.

$$u_{t_2,i} = \frac{\pi_{u,i} \cdot u_{i,t_1}}{\sum\limits_{j=1}^{N_W} \pi_{g,j} \cdot g_{j,t_2}}, \qquad i = 1,...N_D. \tag{5.67}$$

Similarly, $g_{t1,i}$ can be obtained by

$$g_{t_1,i} = \frac{\pi_{g,i} \cdot g_{i,t_2}}{\sum\limits_{j=1}^{N_D} \pi_{u,j} \cdot u_{j,t_1}}, \qquad i = 1,...N_W. \tag{5.68}$$

The P-matrix can finally be determined by applying the elements to (5.57).

5.6.7 Hardware Implementation

Fig. 5.39 shows a simplified block diagram of the hardware implementation. The major aspect hereby is the generation of random state transitions according to the Markov model. Two blocks - "Markov model 1" and "Markov model 2" - are responsible for the noise occurrence and the noise level respectively. They have the same structures, and deploy an inverse transform sampling method to manage the

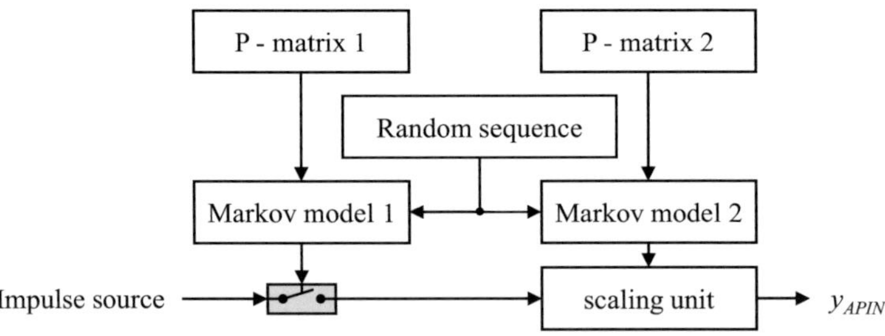

Fig. 5.39 Simplified block diagram for the generation of aperiodic impulsive noise.

state transitions. The inverse transform sampling is a fundamental approach to generate random values according to a pre-defined cumulative distribution function.

In the inverse sampling method, two cumulative probability distribution matrices P_{C1} and P_{C2} are calculated by

$$
P_{cx}(i,j) = \begin{cases} 0, & j=1 \\ \sum_{k=1}^{j-1} P_x(i,k), & i=1,...,n, \;\; j=2,...,n+1 \end{cases} ,x=1,2,
\tag{5.69}
$$

where P_1 and P_2 are predefined transition probability matrixes. P_{C1} and P_{C2} have both n rows and $n+1$ columns and they fulfill

$$
P_{cx}(i,n+1)=1, \quad i=1,...,n, \quad x=1,2.
\tag{5.70}
$$

Uniformly distributed random samples, denoted by R, are generated by the "Random Sequence" block. Each value of R is compared with the elements in the ith row in matrix P_{C1} or P_{C2}. The row index i also refers to the index of states (either impulse-free or impulse state). A system that resides in state i changes to state j only if

$$
p_{i,j} \le R < p_{i,j+1}.
\tag{5.71}
$$

As soon as the process switches to the jth state, the next value of R is compared with each element in the jth row of P_{C1} or P_{C2}. Since a fixed-point algorithm is simpler to be implemented onto a FPGA than a floating-point algorithm, the probabilities are quantized and scaled to 16-bit unsigned numbers. P_C is then predefined as

$$
P_C = \begin{bmatrix} 0 & 65532 & 65532 & 65532 & 65533 & 65535 \\ 0 & 0 & 65532 & 65532 & 65512 & 65535 \\ 0 & 0 & 0 & 64880 & 65081 & 65535 \\ 0 & 88 & 133 & 6554 & 65535 & 65535 \\ 0 & 794 & 1194 & 58982 & 58982 & 65535 \end{bmatrix}.
\tag{5.72}
$$

The first five columns refer to state 1 to state 5, while the last column refers to probability 1 in conventional expression. States 1 to 3 refer to no impulsive noise, while states 4 and 5 represent the occurrence of impulsive noise. It is assumed that a state transition is sampled every 10 µs.

Fig. 5.40 shows an example of the state transition control. As soon as the system is powered on, it enters state 1 and no noise is generated. R gets the value 27330 and it is smaller than $P_C(1,2)$. Therefore, no transition takes place. 10 µs later, the number 65534 is assigned to R which is greater than $P_C(1,5)$. The system enters state 5 at the next 10 µs interval, where the switching signal is asserted. Asserted and non-asserted signals refer to impulse and impulse-free states respectively. The next value for R is 38107 which is smaller than $P_C(5,4)$. Therefore the process jumps to the third state 10 µs later. The same rule applies for the rest of the time. The period of an asserted "1" to the switching signal determines the pulse width, and the span between two such pulses is the time interval in which no impulsive noise appears.

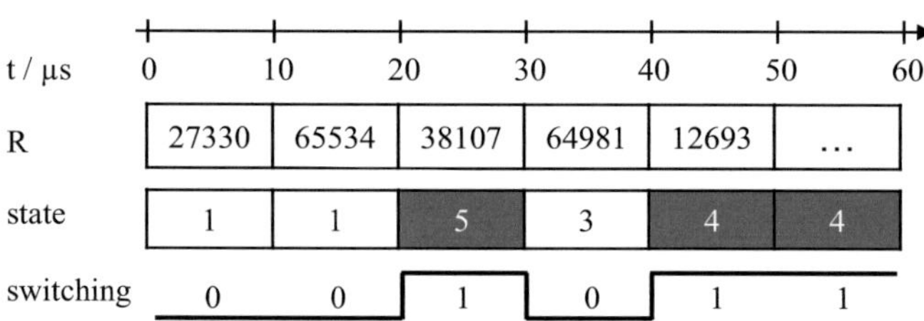

Fig. 5.40 State transition according to the Markov model.

5.7 Burst Noise

The emulation of burst noises is based on a two-level Markov process. Fig. 5.41 shows the structure. All blocks are implemented in the same manner as those for APIN. However, there is no scaling unit. Most burst noise variants are so strong with respect to their amplitude that an ongoing data transmission suffering from such noise will definitely fail. The output of the first Markov model determines the connectivity of the random sequence to the second Markov model.

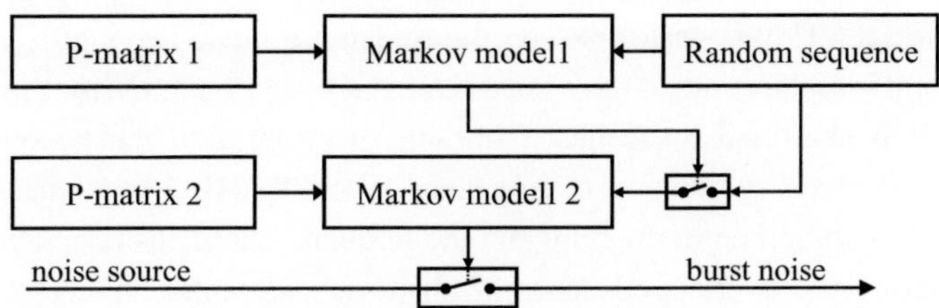

Fig. 5.41 Diagram for emulating burst noise.

Fig. 5.42 illustrates the relationship between the outputs of both Markov models. Burst noise can be generated only if the output of the first model is asserted. The first model defines the noise group, while the second one generates the individual pulses.

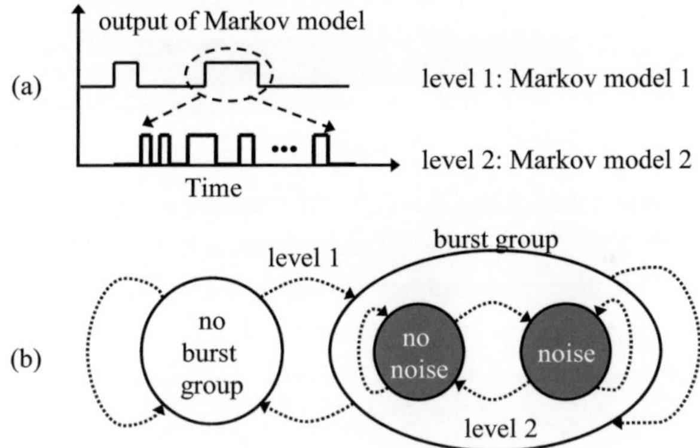

Fig. 5.42 Generation of a switching sequence using a two-level Markov process.

5.8 Colored Background Noise

The colored background noise is a collection of low-level noise types from many unknown sources. Its average power level depends on the number and the types of connected and active electrical devices. Therefore, it can also be considered to have cyclo-stationary characteristics – see e.g. [COR10] and [KAT06]. In order to investigate the time-variance, it is necessary to divide the noise waveform into multiple segments and to estimate the instantaneous PSD of each segment. For this purpose, a STFT is performed on the remaining noise waveform, after narrowband and impulsive noise have been removed - see Fig. 5.33 (b). Furthermore, the colored background noise has a smooth spectrum, and the power spectral density is a decreasing function of frequency [HOO98] [BAU06]. Therefore, the STFT result is smoothed in the time and the frequency domains respectively. Fig. 5.43 (a) shows the overlapped PSDs of all noise segments. The averaged PSD can be approximated by the sum of two exponential functions, i.e.

$$\hat{P}_{BGN}(f) = a \cdot e^{b \cdot f} + c \cdot e^{d \cdot f}, \tag{5.73}$$

where f is frequency in kHz. Table 5.6 shows a set of coefficients which fits with the averaged PSD.

Table 5.6 Coefficients of the Polynomial in (5.73)

a	b	c	d
0.4413	-0.12	$3.132 \cdot 10^{-4}$	$-1.32 \cdot 10^{-4}$

Fig. 5.43 (b) shows the smoothed STFT of the remaining noise. The PSD fluctuation over time is obvious in the frequency range below 100 kHz. Comparing the fluctuation with the mains voltage shown in Fig. 5.33 (b), the maximum of the noise level is synchronous with the peak of the mains voltage. Perturbations can also be observed at higher frequencies. Since the noise level is relatively low, these perturbations are ignored for simplicity.

Let $m(t)$ denote the mains voltage, the parameter $a(t)$ can be obtained by

$$a(t) = \frac{|m(t)|}{A_0} + A_1, \tag{5.74}$$

where A_0 and A_1 determine the scaling factor and the minimum level of the fluctuation respectively. Table 5.7 lists a set of recommended parameters for modeling the time-variant PSD of background noise.

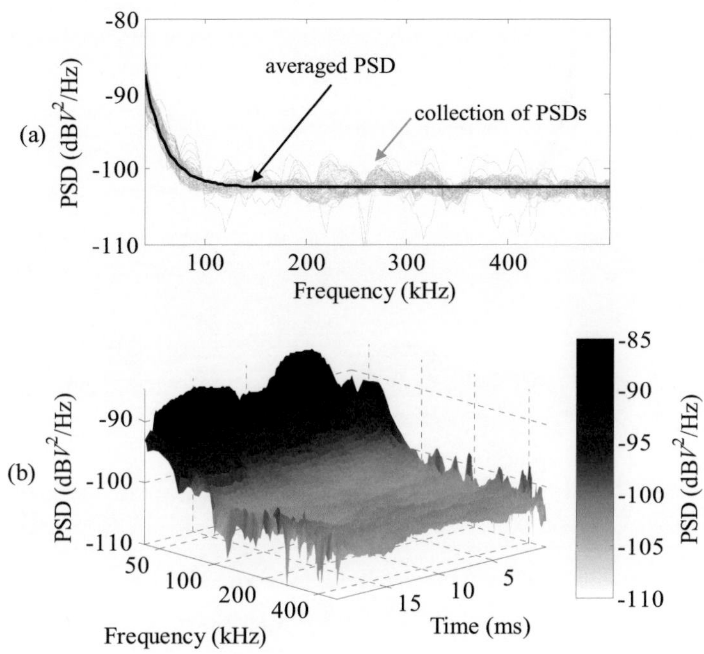

Fig. 5.43 PSD of background noise. (a) PSDs (b) time-frequency view of the PSD.

Table 5.7 Recommended Parameters for Background Noise

A_0	A_1	b	c	d
200	1	-0.12	$3 \cdot 10^{-4}$	$-3 \cdot 10^{-4}$

Fig. 5.44 (a) shows the simulated PSDs, corresponding to the overlapped PSDs in Fig. 5.43(a). Fig. 5.44 (b) is the time-frequency representation of the simulated PSDs.

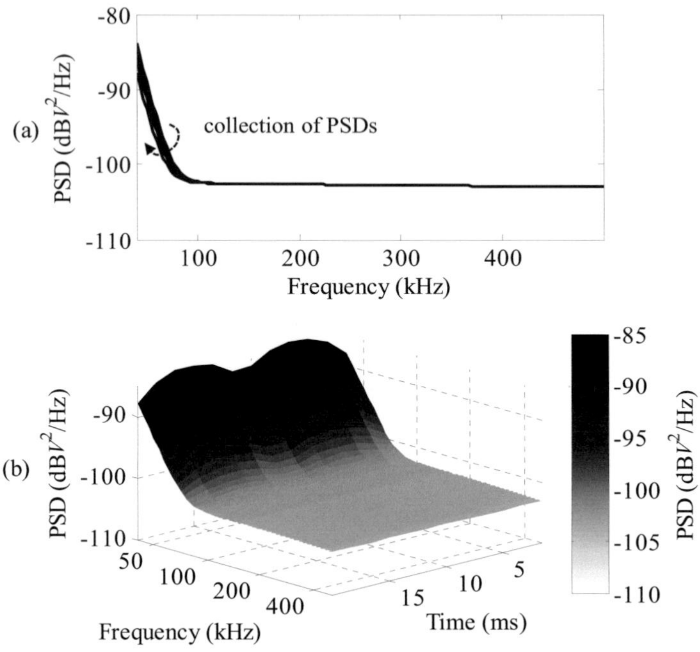

Fig. 5.44 Emulated PSD of background noise. (a) PSD profiles (b) time-frequency representation of the PSD.

6 Evaluation of NB-PLC Systems

In order to illustrate the flexibility and reliable performance of the emulator-based testbed, several realistic examples are presented in this section.

6.1 System Evaluation under Real Channel Conditions[4]

This example verifies whether and how well the system performance in the real-world channels matches the test results from the proposed emulation platform. A small low voltage grid on a university campus is chosen for this purpose. The grid is a three-phase 230V/50Hz system, and all the investigations are made on the same phase in the CENELEC A-band (3-95 kHz). Fig. 6.1 shows the grid structure and three measurement locations. The grid is composed of four-wire underground cables (three phases plus neutral). S1, S2 and S3 denote three measurement points: a transformer station, an office building and a workshop building respectively. S2 is connected to S1 via a ring which supplies seven buildings in total. S3 is connected via a branch from the office building.

In the first step, relevant channel characteristics such as the attenuation and the noise scenarios are measured. Data transmissions are made over the same channel between two measurements using two OFDM-based PLC systems in intervals between two adjacent measurements. The BER for each data transmission is recorded as an assessment of the link quality. In the second step, the channel properties are characterized and reproduced using a PLC channel emulator. The same PLC systems are connected to the emulator. The same data transmissions are made again and the BER values are measured as well. In the last step, the accuracy of the channel emulation and the validity of the emulator-based performance evaluation are verified by comparing the BERs measured under real channel conditions with the BERs obtained with the emulated channel conditions.

---

[4] Parts of this section have been published in the Journal paper [LIU114]: W. Liu, M. Sigle and K. Dostert, "Channel characterization and system verification for narrowband power line communication in smart grid applications," IEEE Commun. Mag., vol. 49, pp. 28-35, Dec. 2011.

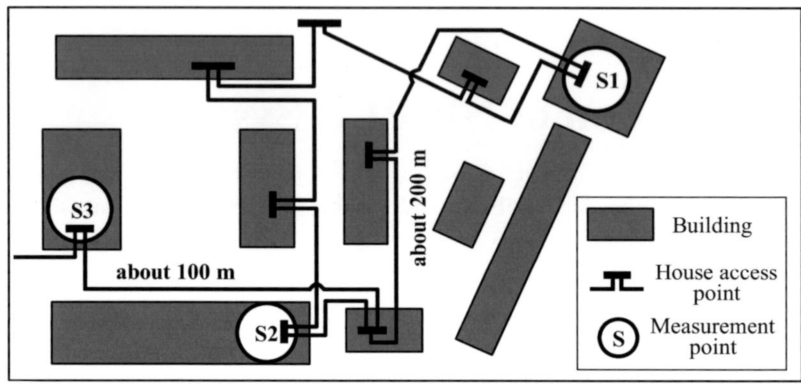

Fig. 6.1 Campus grid under investigation.

6.1.1 Measurement Procedure

In order to obtain all relevant parameters for the channel emulation, a systematic measurement method has to be applied. Three highly flexible measurement and communication platforms – as described in [SIG11] – have been deployed, one platform for each measurement location. Each platform allows coupling arbitrary test signals into the powerline, as well as receiving and capturing signals of arbitrary duration. In addition, the platforms can also be configured as OFDM modems for data transmission. The measurement campaign is composed of many measurement cycles. During each cycle a complete channel investigation is made. In order to study the time variance, the measurement cycle repeats every ten minutes and a measurement procedure lasts for about 2 hours.

In the first step, all three platforms record the local noise at the same time for 30 seconds. The second step measures the transfer functions from one location to another. Since the channel transfer characteristics feature usually no symmetry (for example, the direction from S1 to S2 behaves different from the direction from S2 to S1), they must be measured for all possible directions. As a result there are totally 6 directions to investigate. One of the three platforms is allowed to transmit test signals while the other two must record their received signals for the estimation of channel distortions. This way always two directions are investigated at the same time. When one platform finishes its transmission, it switches into the receiving mode, and another platform starts to transmit test signals. This process continues until the attenuation values are captured for all 6 directions. In the last step, the OFDM module of each platform is activated, data frames instead of test signals are transferred, according to the same sequence and the BER values are recorded as an assessment of the instantaneous link quality.

6.1.2 Emulating Channel Transfer Functions

The measured curve is interpolated first, so that the frequency resolution of the measurement can be matched to that of the emulation. The reproduced transfer functions of the investigated data links are shown in Fig. 6.2. The lines with asterisks correspond to the measured attenuation values, and the solid curves are the emulated CTFs. There is an error of 1.5 dB at 30 kHz for the link from S1 to S3. However, its influence is negligible, since it is relatively small compared with the attenuation value (more than 50 dB) at this frequency, and it lies outside of the transmission band. The coupling loss is relatively small since the transmitter of the platform has very low output impedance. Therefore, its influence is neglected and the access impedance is not emulated here.

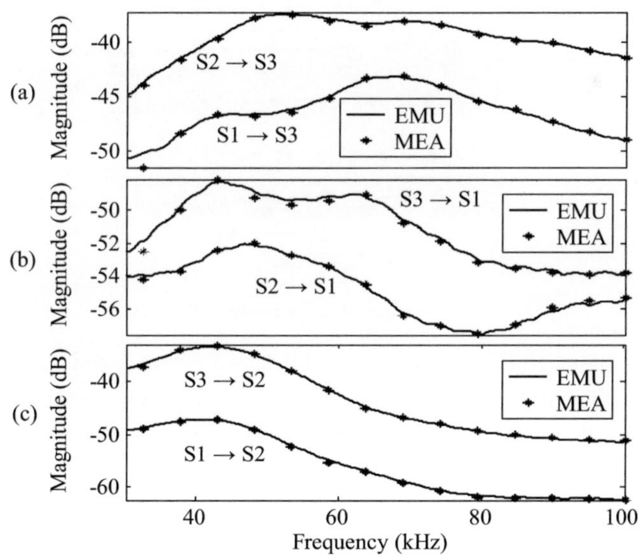

Fig. 6.2 Comparison of measured and emulated transfer functions.

6.1.3 Emulating Noise Scenarios

For the noise emulation we take the scenario measured at S2 as an example. Since the data transmission is made between 60 to 100 kHz, the noise is analyzed and emulated for this frequency range. Due to the time-varying and frequency-selective nature, the analysis is performed in time and frequency domains using STFT.

Fig. 6.3 shows waveform and PSD of the measured noise. Narrowband interferers with periodical fading can be observed at around 65.5, 67.5, 71, 76.5 and 88 kHz in plot (b). All the fading seems to be synchronous with mains voltage and appears every 10 ms. A total of 12 narrow spectral peaks are found within 40 ms. They are the spectral components of broadband impulsive noises. By applying the analysis approach introduced in chapter 1, these impulses can be divided into three groups.

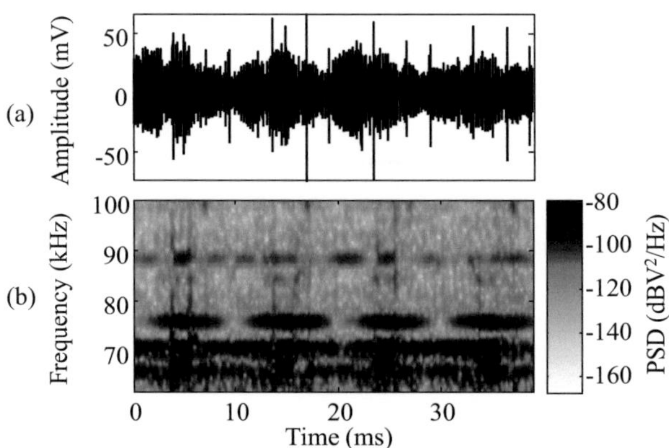

Fig. 6.3 Measured noise scenario for S2.

The group index and the arrival times of the individual impulses are shown in Table 6.1. It is obvious that the impulse in each group has a repetition rate of 100 Hz. Last but not least, the colored background noise has weak time varying feature in this frequency range. It can be modeled by a time-invariant exponential function of the frequency. By applying the emulation approaches for each noise type, the overall noise scenario is reproduced for S2. As shown in Fig. 6.4, the emulated noise scenario retains the essential characteristics.

Table 6.1 Arrival Time and Group Index

Group Index	Arrival time (ms)
1	3.3, 13.3, 23.3, 33.3
2	5.5, 15.5, 25.5, 35.5
3	6.8, 16.8, 26.8, 36.8

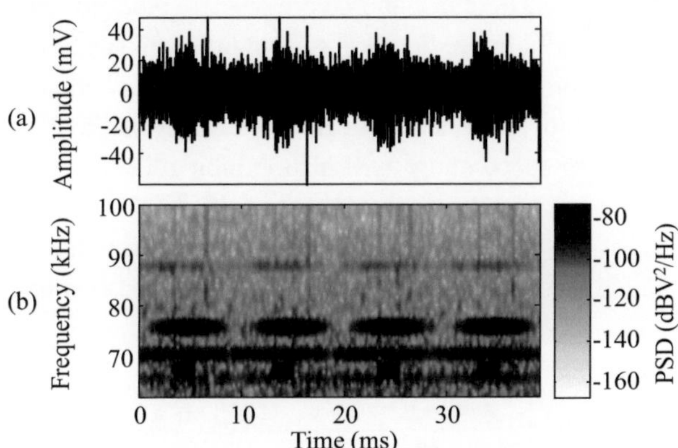

Fig. 6.4 Emulated noise scenario for S2.

6.1.4 Device under Test

Besides the channel measurement facilities, the FPGA used for the measurement platform also hosts a flexible and easily adaptable OFDM modem core. It consists of independently configurable blocks. The modem core supports arbitrary carrier utilization, different modulation schemes and different methods of synchronization. Also preambles and frame length can be selected. Thus, a variety of multi-carrier systems can be configured and tested with low effort and complexity. A simple OFDM realization has been configured on the FPGA for the following tests. The system is using 48 carriers, ranging from 79 kHz to 95 kHz. Each carrier is modulated with DBPSK for robust communication. Symbol synchronization is performed by using the zero-crossings of the mains voltage. Thus, the total symbol duration – including a guard interval – is chosen as 1/6 of the mains period, i.e. 3.3 ms in a 50 Hz environment. The frame length is 9 OFDM symbols. In the field measurements 200 frames with random data are transmitted.

6.1.5 Test Results

Fig. 6.5 illustrates the BER results recorded during the channel measurement and with the emulator-based test platform. The link S3-S2 exhibits the best communication quality. While the links S1-S2, S2-S1 and S2-S1 show very high BER values. The BER results obtained by using the test platform are very close to those of the field tests. Therefore, the laboratory test platform can obviously be regarded as highly useful to predict PLC system performance in real-world channels. We can also reproduce worst-case channel conditions obtained from measurements, and apply them for testing different PLC systems. Systems that pass such worst-case tests successfully will also exhibit reliable and robust performance in the real-world applications. Therefore the emulation environment proposed in this thesis can be regarded as a valuable means to reduce the complexity and the time which is usually needed for selecting a best PLC solution for a given application.

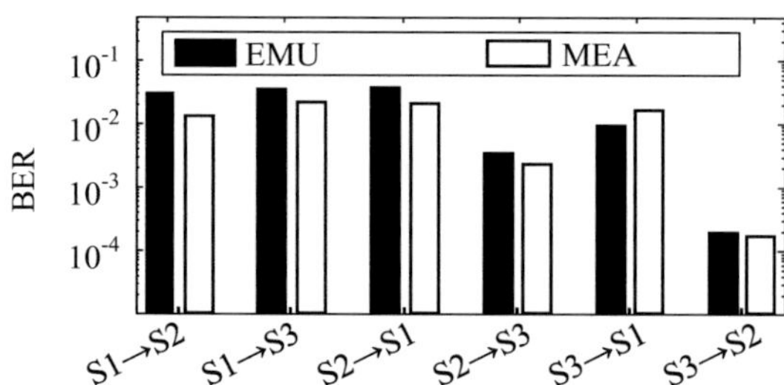

Fig. 6.5 Comparison of BER results between field test and test made at the emulator.

6.2 Margin Test using Synthesized Conditions

In this test step, the performance of DUTs is investigated under synthesized channel conditions. The goal is to find out the signal to noise ratio margins (SNRM) required by a PLC system to maintain a defined performance under specific channel conditions. The test is usually carried out by changing the channel conditions gradually until the data link breaks. A lower margin indicates higher robustness and better immunity. This test is helpful to identify the specific channel impairment which degrades the system performance significantly. As a result, clear hints towards performance improvement can be obtained. Furthermore, since different PLC systems usually show different margin values, the test results must be considered to be system-specific. Therefore, synthetic benchmarking can be performed for systems from different manufacturers by measuring and comparing the corresponding SNRM parameters. The synthesized channel impairments can be noise patterns extracted from measured noise scenarios, or they can be created artificially. This test methodology doesn't need a large amount of preliminary measurements. It is especially suitable for functional verification during a PLC system development. It is worth mentioning that the individual synthesized impairments cannot reflect the situation in the real powerline networks completely. Therefore, each individual test cannot deliver an accurate evaluation of the DUT alone. However, carefully defined test patterns and test procedures can still reveal the actual performance of the DUTs, and a comprehensive performance evaluation can still be obtained by considering all test results.

6.2.1 Typical Margin Test Routine

Generally the DUTs should be evaluated with tens of thousands of data packets, so that statistically significant results can be achieved also for low bit error rates. Since in practice the data rate of the DUTs used here is relatively low, a single test period usually has to last for several hours. For a test case composed of dozens of periods, the process can even take several days. It is necessary to automate the process, so that test cases can run day and night, and the parameters can be changed without action of test personnel. Fig. 6.6 gives an example of such a typical test routine. The first step is to configure the DUTs, either as transmitters or as receivers. The emulator must be updated with a channel transfer function in the next step. Then the selected attenuation values are determined and saved together with the control code in a look-up table. A test period begins with the configuration of the SNR value, and all current configurations are saved. As soon as a frame of stress pattern (predefined combinations of payload data used for tests) is launched, a timer is started for timeout counting.

If the receiver has obtained the stress pattern and returned it to the test server before timeout, the frame is considered to be received successfully, and it is used to calculate the BER and frame error rate (FER). Otherwise a frame loss event is

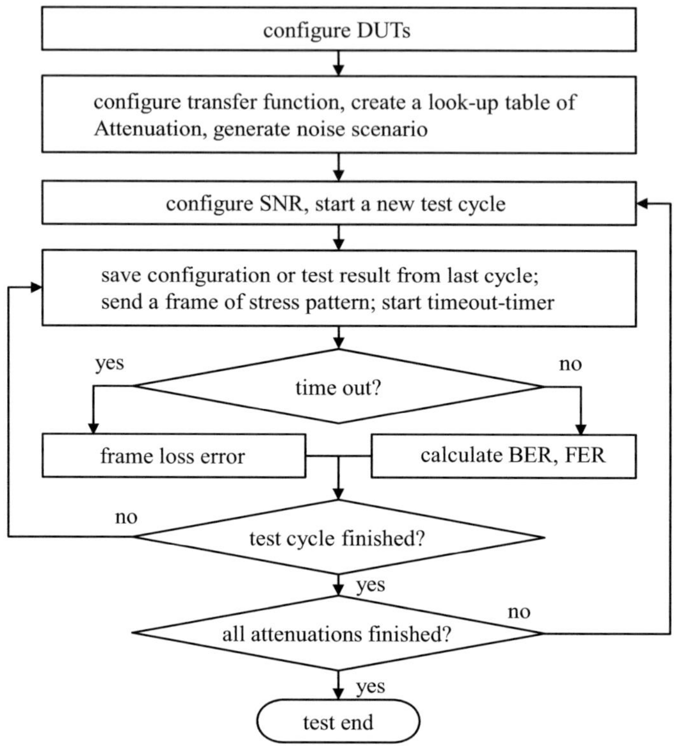

Fig. 6.6 Flow chart of a typical margin test routine.

registered. The process repeats until the test server has run all stress patterns for the current test cycle. The next cycle starts with a new SNR value. After all SNR values to be considered have been applied, the whole test is completed. The following parts of this paper will focus on the emulation of the channel transfer function and the determination of the look-up table which are needed for the second step.

6.2.2 Case Study: Frequency-Selective Attenuation

To investigate the influence of frequency-selective attenuation on the noise margin of different DUTs we apply additive white Gaussian noise to the transmitted signal and keep the noise PSD constant throughout the test process. The transmitted signals of DUTs are attenuated step by step, until no data link can be established any more. The BER and the frame loss rate (FLR) are then used to evaluate the performance of a DUT. Table 6.2 lists a selection of involved DUTs from different manufacturers. DF1 and DF2 are both based on S-FSK technique. DO3 and DO4 are based on OFDM and are subject to the same PHY - layer and MAC - layer specifications.

Table 6.2 DUTs in the Proposed Margin Test

DUT	DF1	DF2	DO3	DO4
Technology	S-FSK	S-FSK	OFDM	OFDM

Four channels are involved in the noise margin test: a flat channel (CF1) and three channels with frequency-selective attenuation (CS2, CS3 and CS4). CS2 and CS3 are chosen from measurements reported in [DOS00]. The amplitude responses of them are illustrated in Fig. 6.7. Both have strong notches in common. Sharp increase of attenuation between 20 and 40 kHz is also observed [DOS00]. CS4 has a similar course as CS2, however, the attenuation at 74.5 kHz is more than 24 dB higher than that at 60 kHz.

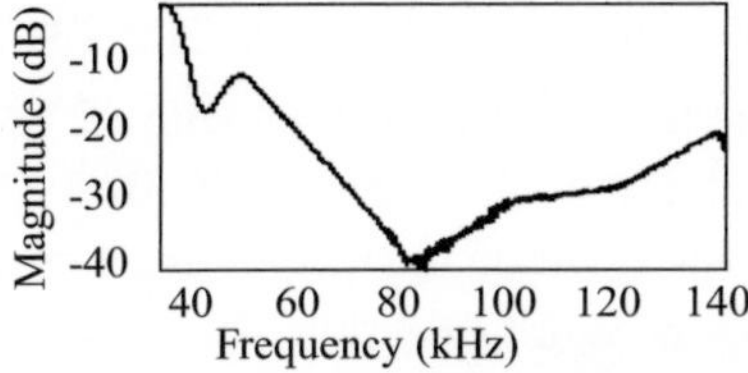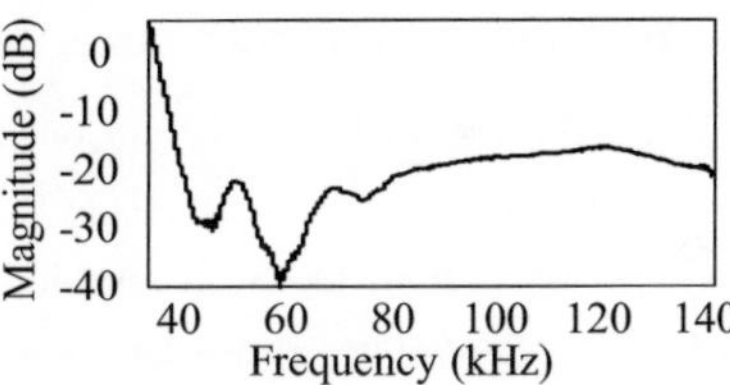

Fig. 6.7 Magnitude courses of CS2 and CS3.

To compare the robustness of OFDM with S-FSK, we have tested DF2, DO3 and DO4 with CF1, CS2, and CS3. The impairments caused by different channels on the implementation can also be observed. To make the two techniques - the S-FSK and the OFDM - comparable, we control the transmitted signals in such way that the PSDs of the signals "seen" by the receiver have the same maximal value. Fig. 6.8 shows two examples. In the first plot the channel condition CS2 is emulated, while in the second plot CS3 is involved. The white noise has a PSD of -76 dBW/Hz. The PSD curves of DO3 and DO4 are almost completely overlapping. DF2 has two carriers, one at 60 kHz, and another around 75 kHz.

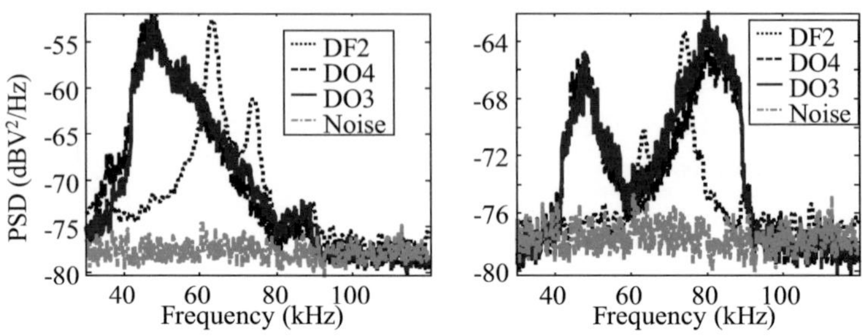

Fig. 6.8 PSD courses of signals distorted by CS2 and CS3.

The PSD value of the 60 kHz-carrier is the same as the peak value of DO3 and DO4. Apparently, the signal power injected by DF2 is lower than that of DO3 or DO4 in this case. CS4 is employed to check the implementation of the quality measurement mechanism according to [IEC01] in DF1 and DF2.

The transmitters launch signals of the same level, as shown by "DF1-Tx" and "DF2-Tx" in Fig. 6.9. "DF1-Rx220" and "DF2-Rx220" are the measured PSDs of the signals at the inputs of the receivers, when the transmitted signals are attenuated. Again, both received signals have the same PSD curves. However, only the carrier at 60 kHz is still visible, the other one at 74.5 kHz has disappeared.

Fig. 6.10 through Fig. 6.12 show the results of tests carried out with CF1, CS2 and CS3 respectively. Obviously, DF2 exhibits a larger noise margin (over 13 dB) than DO3 and DO4. DO3 obtains lower BER values than DO4 under all channel conditions. However, the FLR values show significant diversities. DO3 has slightly lower FLR values under CF1, but almost the same values under CS3, as shown in Fig. 6.10 and Fig. 6.12 respectively.

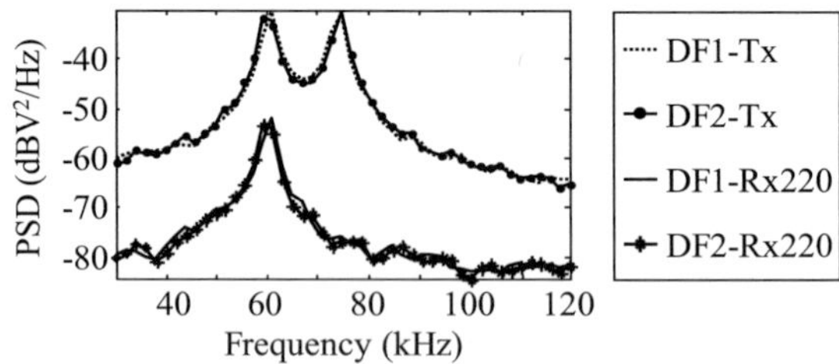

Fig. 6.9 PSD courses of signals of DF1 and DF2.

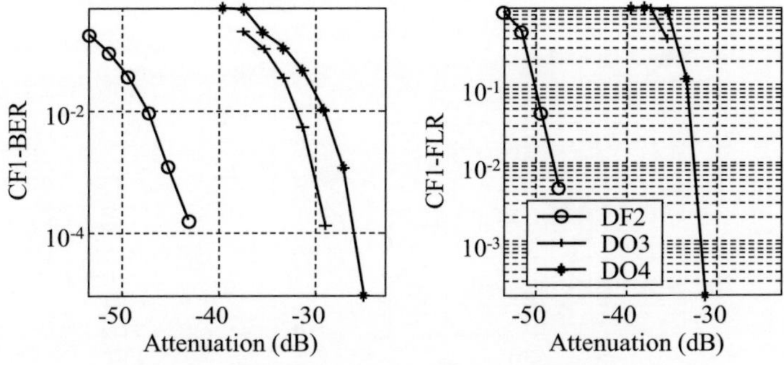

Fig. 6.10 BER and FLR with channel condition CF1.

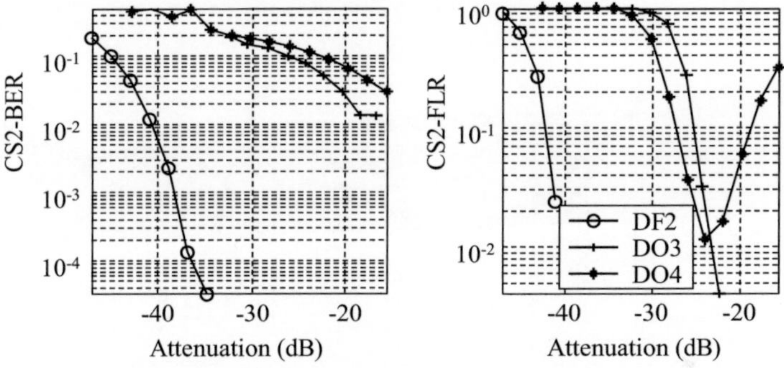

Fig. 6.11 BER and FLR with channel condition CS2.

In Fig. 6.11 DO4 exhibits lower FLR values than DO3, when the attenuation of the transmitted signal is higher than 24.5 dB. Before the attenuation reaches that value, DO4 loses surprisingly more frames, even though the attenuation decreases.

Fig. 6.13 shows the BER and FLR curves of DF1 and DF2 with channel CS4. DF2 begins to suffer from bit errors and frame losses, when its transmitted signal is attenuated by 43.8 dB. The BER and FLR rise as the attenuation increases until not a single frame can be detected any more at 58 dB. DF1 seems to have a difficulty with CS4, i.e. the BER as well as the FLR are relatively high, even when the attenuation is quite low. The error rate is so high that no useful data can be transferred any more. It is evident that the quality measurement mechanism - as introduced in 2.2.1.2 - in DF1 is either bad, or not implemented.

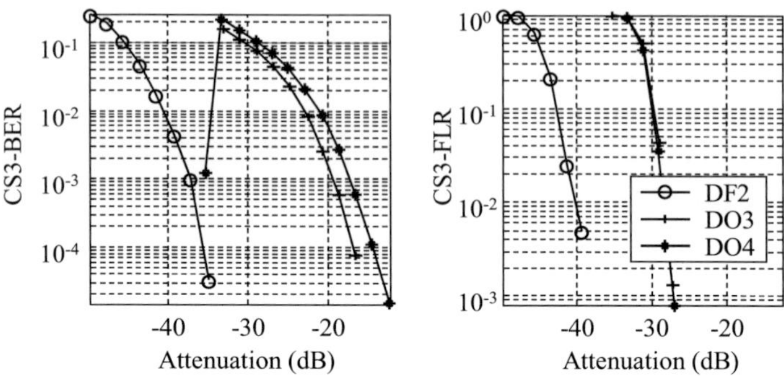

Fig. 6.12 BER and FLR with channel condition CS3.

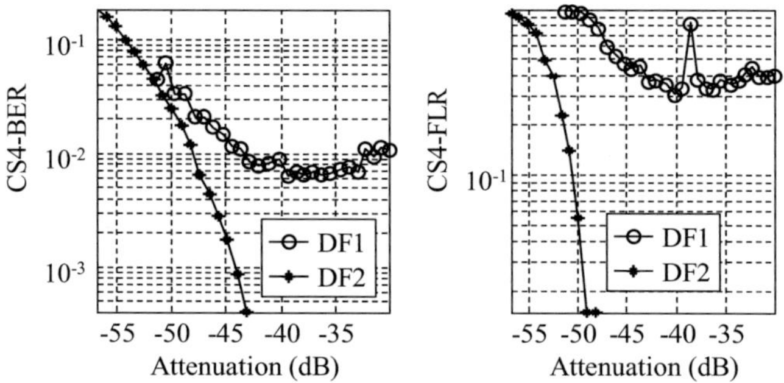

Fig. 6.13 BER and FLR with channel condition CS4.

6.2.3 Case Study: Influence of Phase Distortions[5]

This section investigates the noise margin of a DCSK system under phase distortions. DCSK is a spread spectrum modulation technique. In principle, a chirp signal and its cyclically rotated waveform are used to carry digital bit patterns. The receiver employs correlation for the symbol detection and the demodulation. The original chirp waveform is stored as a template. In the demodulation process the template and its cyclic rotations are used as parameters of the receiver's -

---

[5] Most contents in this part have been published in [LIU12]: W. Liu, M. Sigle and K. Dostert, "Channel phase distortion and its influence on PLC systems," in Proc. IEEE Int. Symp. Power Line Commun. Applicat., Beijing, China, Mar. 2012, pp. 268-273.

178

correlator. There is a correlation result for each rotation, and the one with the largest value is considered as the result of a symbol detection. Finally the symbol is demodulated to obtain the bit pattern. For transmissions in the CENELEC A band, the same chirp waveform is repeated in three sub-bands: 18-44 kHz, 38-63 kHz and 58-89 kHz [JOE09].

In total, four different phase responses are emulated and their corresponding group delay profiles are shown in Fig. 6.14. The first phase response is linear with an almost constant group delay. The test result with this linear phase channel is considered as a reference. Comparing this result with those obtained under non-linear phase distortions gives a straightforward assessment of performance degradation. The second channel has random phase shifts from a uniform distribution. The delay values are distributed randomly, and a sharp drop appears at 22 kHz. For the third channel we use repeated patterns in the sub-bands 28-39 kHz, 43-54 kHz and 67-78 kHz for the purpose of covering all three transmission bands of the DCSK system. For the last channel two more patterns are added, so that we can see whether the performance becomes worse with an increased non-linear phase distortion or not.

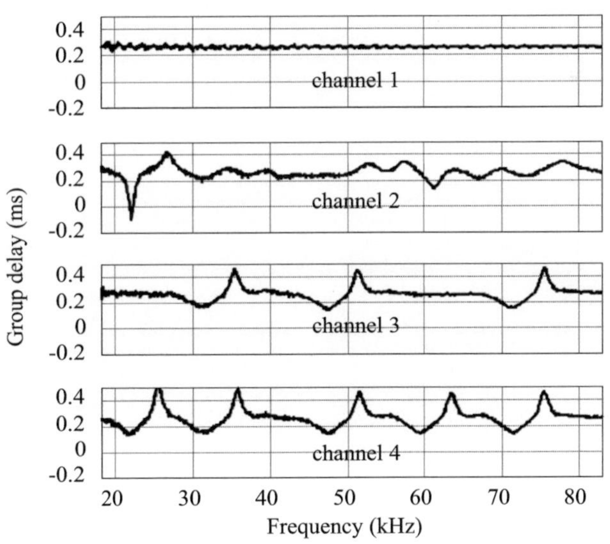

Fig. 6.14 Emulated group delay profiles.

In order to investigate the influence of non-linear phase distortion alone, we keep the other channel conditions as simple as possible. We match the amplitude responses to obtain almost flat PSDs of received signals in all channels (rx1 through rx4), as shown in Fig. 6.15.

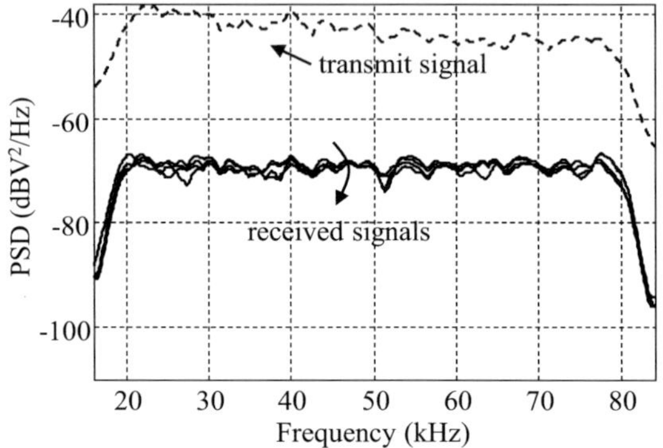

Fig. 6.15 PSD of transmitted and received signals.

For the noise scenario, the AWGN instead of other typical PLC interference is emulated because AWGN has flat PSD and it has statistically equivalent influence on all frequencies. As shown in Fig. 6.16 the PSD of the AWGN is kept constant during all tests, and the transmitted signal is attenuated step by step, until the data transfer stops.

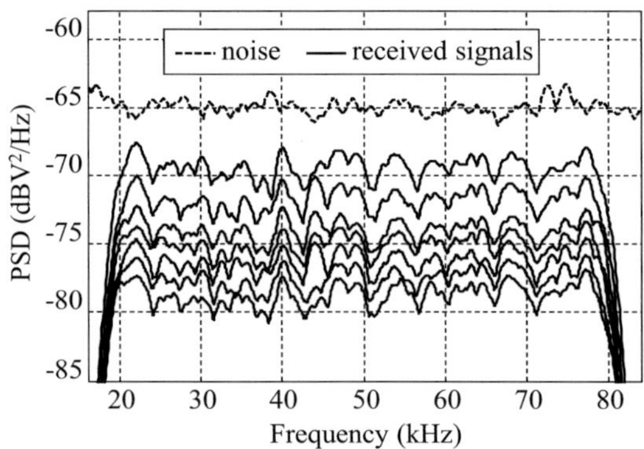

Fig. 6.16 PSD of noise and attenuated received signals.

Two DCSK receivers from the same manufacturer are tested with the emulated channels. The PLR over SNR plot is shown in Fig. 6.17. The dotted lines are for the first receiver while the solid lines are for the second one. Obviously, both receivers have different noise margins. The first receiver is more sensitive than the second one, in particular under the first and last channel conditions. It can also be seen that the receivers show the same tendency. They both have larger noise margins under linear phase conditions (curve 1-1 and 2-1). The margins are reduced by about 4 dB when the phase response becomes non-linear and random (1-2 and 2-2). Increased phase distortion further reduces the noise margin (i-3 and i-4, i = 1, 2).

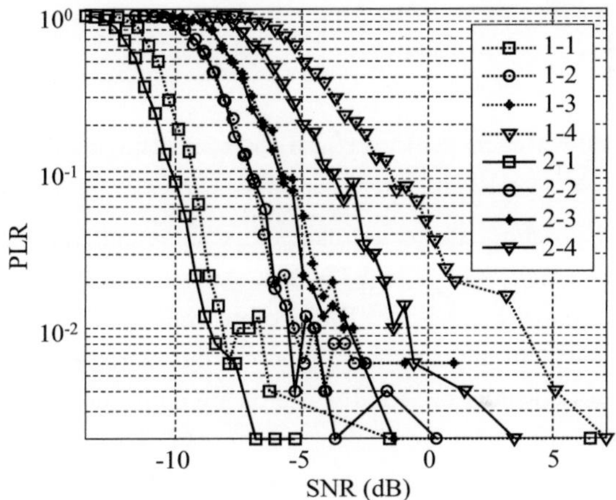

Fig. 6.17 PLR vs. SNR. The legend "i-j" means the ith DUT is tested with j^{th} group delay profile.

This margin test has highlighted the influence of non-linear channel phase distortion on correlation results. This distortion degrades the quality of frequency domain phase modulation and DPSK. Matched-filter-based synchronization and modulation can also be affected. Channel estimation and phase compensation should be implemented to improve the robustness.

6.2.4 Case Study: Margin against Periodic Impulsive Noise

For an impulsive noise test, periodic impulsive noise with amplitude A_M, center frequency F_M and pulse width T_F will be used. The meaning of the parameters is explained in Fig. 6.18. Table 6.3 lists the values of the parameters. The channel has flat attenuation, and the attenuation of the transmitted signal is increased step by step, until no communication is possible any more. Since the time interval between two adjacent symbols is not constant, it is hard to estimate the energy ratio between the signal and the PIN noise. Therefore, the abscissa in the following PLR plot shows the attenuation value instead of the SNR.

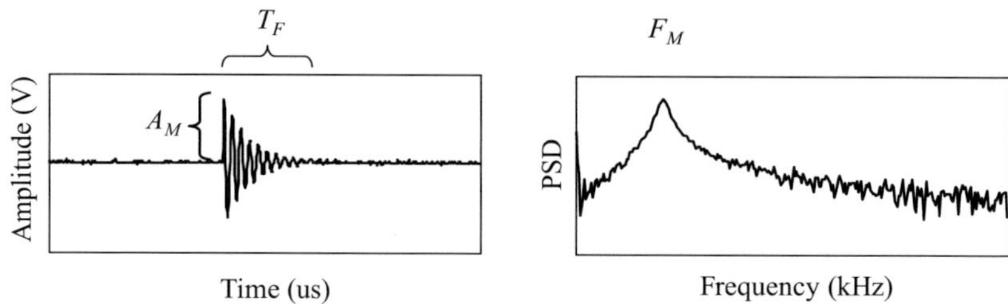

Fig. 6.18 Waveform (left) and PSD (right) of emulated impulsive noise.

In total 6 PLC systems are tested using this noise pattern. Modulation of the DUTs includes DCSK, S-FSK and OFDM. Among the OFDM-based modems there are a G3 compatible, two PRIME compatible systems from two different providers, and a system which is used in street lighting applications. In this test, the data rate is not considered and all systems are configured in their most robust modes. The measured PLRs are shown in Fig. 6.19. The attenuation value is the ratio of the received signal level to the maximum transmission level that the system can deliver without violating the CENELEC norm. Generally speaking, sin-

Table 6.3 Parameters of Impulsive Noise

Index	Repetition rate / Hz	A_M/ V	T_F / µs	F_M/ kHz
PIN	100	1.02	5000	65

gle carrier technologies are more robust than multi-carrier technologies. DCSK has the highest immunity against the noise. The S-FSK system is more sensitive to this noise pattern than the DCSK system. It needs a more than 20 dB better SNR value than the DCSK system for the PLR not to exceed 10^{-2}. The G3 compatible system seems to have the lowest margin to keep the PLR below 10^{-1}. The

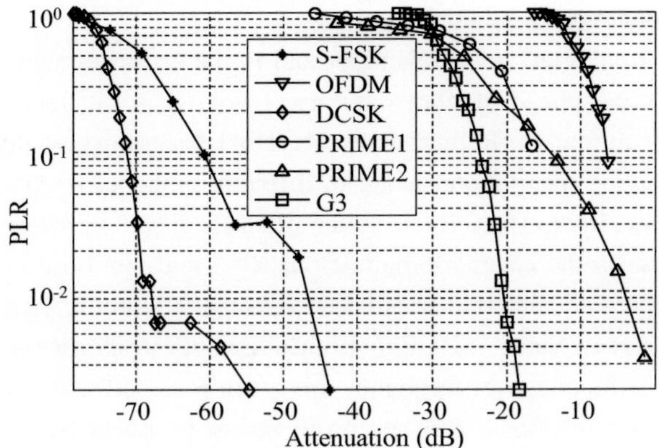

Fig. 6.19 Noise margins of different PLC systems.

two PRIME compatible systems have similar noise margins. The last OFDM-based system needs the highest noise margin to reach similar performance.

6.3 Reference Channels

In typical benchmarking, it is desirable to run tests under reference channel conditions. These channels shall be realistic and representative; thus the definition of such reference channels should be based on statistically reliable samples taken from a sufficiently large measurement database.

Extended measurement campaigns have not been in the focus of interest for this thesis. It is not realistic to investigate all possible NB-PLC channels all over the world for the definition of reference channels. Instead, a few own measurements performed within this thesis and representative results reported by other researchers are combined and analyzed. As an outcome, reference conditions are proposed for both the CTF and the noise scenario. Detailed information can be found in appendix A and B respectively.

6.3.1 Reference CTFs

The reference CTFs are divided into two major groups: unloaded and loaded channels. The unloaded channels refer to attenuation caused mainly by cable materials and network topologies. They have smooth curves in the frequency range of interest. Four types – Hb-1 through Hb-4 - are proposed for this group. They cover wiring topologies, both for indoor and in the access domain. Furthermore, in-phase and cross-phase situations are also involved. All of them can find their original molds in real-world measurements. Hb-1 has a lowpass character, and it refers to an indoor channel consisting of different branches [BAU05]. Hb-2 also shows a lowpass filter effect. It has lower attenuation and represents a link on the same LV phase in the access domain. Both Hb-3 and Hb-4 are indoor channels showing highpass characteristics. Hb-3 is crossing different branches while Hd-4 is within the same branch [SUG08]. Both magnitude responses and their corresponding linear functions of frequency are given to describe the unloaded channels. Their phase responses can be considered to be linear over the whole frequency range.

For the loaded channel group, there is no significant difference between indoor channels and access domain. The CTFs are mainly characterized by a number of local extremes. The number and the dynamic range of the extremes reflect the complexity of the loading conditions. This group also contains four channel types denoted as Hd-1 through Hd-4. The first three types refer to the most typical European indoor and outdoor channels. They find their original molds in the three typical reference models proposed in [BAU06]. These models are defined out of the statistical analysis of the extreme values in the amplitude responses. The last one has many sharp notches. It represents the situation in which loads with sharp resonances are connected. This kind of notch-rich channels is typical for cross-transformer channels such as those reported in [MAR122]. Originally, the original molds are given in the form of amplitude responses. The resonance-circuit-based models and the algorithm introduced in 0 are applied to get the complex-valued CTFs. Both the amplitude and the phase responses are plotted. Circuit parameters are also given for each single stage of the models.

6.3.2 Reference Noise Scenario

The proposed reference noise scenario is organized by noise classes. In each class, three levels - highly noisy, moderately noisy and quiet - are defined according to different disturbance intensities. For the colored background noise, the model introduced in section 5.8 is used to describe the time-varying PSDs. The averaged PSD profiles are created according to the statistical analysis reported in [BAU06].

They refer to the noise levels with the maximum, 50%, and the minimum cumulative probabilities respectively.

For the narrowband interferers, parameters such as center frequency, bandwidth and maximum noise level are given for individual interferers. Time-varying envelopes are also modeled at frequencies where the noise level fluctuates frequently. They can be applied in highly noisy and moderate scenarios. An occurrence of swept-frequency noise is rarely reported in literature. Therefore, all information is based on own measurements performed within this thesis. Despite the limited database for the SFN, the proposed patterns can still cover most typical indoor and access domain channels.

For the periodic impulsive noise, the oscillations with exponential decay - handled in section 5.6.2 - are used as time-domain waveforms. Noise features, such as repetition rate, amplitude, pulse width, center frequency, as well as delay with respect to mains zero-crossings are listed. Aperiodic impulsive noise and burst noise exhibit random waveforms, since the inter-arrival time and the pulse duration play more important roles than the pure waveform course. The timing behavior is emulated by 5-state Markov models. The sampling period for the noise synthesis is set to 50 µs for a compromise between time resolution and implementation complexity. Quantization is made with 24 bits to achieve an acceptable accuracy. There are 11 parameters for the noise synthesis and their influences on the noise characteristics are dependent on each other. Therefore it is reasonable to keep some of these parameters constant and to modify the others to achieve the noise diversity. For this purpose, parameters such as the average impulse distances $\bar{t}_{D,j}$ ($j=1,2,3$) and the average impulse widths $\bar{t}_{W,j}$ ($j=1,2$) are kept for all three levels, while the relative disturbance ratio r_D and the stationary distributions are defined differently for each noise level. As a consequence, each state of the Markov model has the same average duration for all three noise levels, while the transition probabilities from one state to the other are different. It's worth mentioning that the statistical parameters have been properly scaled, so that these noise types appear more often than they do at real-world channels. Although this modification makes the proposal less realistic, it shortens the test period and makes the test more efficient.

7 Conclusions

In this thesis, a channel emulator has been developed for NB-PLC channels in LV mains networks. The channel emulation is based on a three-stage equivalent electrical model and using the linear periodically time-variant (LPTV) as well as Markov chain to describe the short- and long-term timing behavior respectively. A testbed has been built by using this emulator for the evaluation of bidirectional data transmission on the physical (PHY) layer. Besides, effort has been made with the following aspects:

7.1 Bidirectional Channel Emulation

The bidirectional testbed utilizes two unidirectional emulators, one for each direction. Digital filters are applied to cancel the unwanted echo effect which is caused by the closed-loop formed by two unidirectional emulators. The residual cancellation error could be caused by the estimation error of the impulse responses as well as the quantization error in the digital implementation.

7.2 "Pure" Mains Voltage for Testbed

A UPS is used to isolate the testbed from the real-world mains network, and to provide an alternative supply voltage. LISNs are used to filter high frequency noise coming from the UPS, and to prevent the transmitted signal from reaching the receiver via the mains path. The performance of a commercially available LISN and two specific LISNs is compared. Generally speaking, the specific LISNs have better filtering characteristics than the commercial one. However, in one specific LISN using ferrite-core inductors, undesired impulsive noises may occur, due to the nonlinear nature of ferrite core materials. In the air-core inductor based LISN, non-linearity due to material saturation will not appear. It is most suitable for this application.

7.3 Channel Transfer Function

A measurement platform has been developed to investigate the complex-valued channel transfer functions for bidirectional NB-PLC channels. The platform uses a GPS receiver for accurate time synchronization. A discrete double-frequency approach is used to obtain the amplitude response and the group delay. The CTF for one direction is different from the CTF for the other one, indicating un-symmetry with respect to the channel transmission characteristics. Indoor channels and channels in the access domain exhibit different degrees of un-symmetry. The difference can be explained by the physical lengths of these channels. For indoor channels, the cable lengths are usually in the range between 10 to 50 meters. Such cables cannot decouple the connected electric appliances from the measurement locations. The channel characteristics are determined by the total number of connected appliances. On the contrary, in the access domain, a decoupling effect can be observed due to relatively long cables. The frequency selectivity of the CTFs is mainly "shaped" by the cable topology and the appliances that are close to the measurement locations. Since different locations can have quite different local appliance scenarios, the two directions of a link usually exhibit different CTFs despite of the same network topology they are sharing. In addition to the un-symmetry, the group delay is not constant for both directions. Thus, the channel has nonlinear phase responses. Both the phase response and the group delay show periodic time-varying features. The periodic fluctuations occur mostly below 300 kHz.

A model of cascaded resonance circuits is applied to describe measured CTFs. This model is based on the assumption that the maximum and minimum extremes in the magnitude responses are caused by parallel resonant circuits (PRCs) and series resonant circuit (SRCs) respectively. An algorithm has been proposed to decompose an arbitrary amplitude response into multiple PRCs and SRCs. The complex-valued CTF is then obtained by cascading all resonant circuits and calculating the overall transfer function. With this model, both the phase response and the group delay can be estimated from an amplitude response with acceptable accuracy.

Digital FIR filters are utilized to emulate the channel transfer function. An extended frequency sampling algorithm provides a flexible and straightforward way to design such FIR filters for emulating any customized complex-valued CTF. In addition, recommendations are given to reduce the redundancy of the implemented filters, so that requirements for real-time applications can be met.

7.4 Noise Scenario

A Gaussian process is used to generate colored background noise. This is an important starting point to obtain more complicated random processes. Therefore, the digital generation of Gaussian random numbers has been studied in detail in this thesis. Some common methods such as the central limit theorem (CLT), the Box-Muller, Polar and direct rejection methods are compared. The CLT method is simple to realize, and the quality of generated noise is satisfying. However, it has low convergence speed and low flexibility. The direct rejection method is straightforward, flexible and able to generate noise with arbitrary distribution. However, it exhibits very low efficiency. It is therefore not suitable for high speed applications. The Box-Muller and the polar method are recommended for the current application. The impacts of analog components, such as the reconstruction filter, the coupling circuits, and suppressor diodes have also been investigated. The linear filtering of the noise will not change the Gaussian distribution. However, improperly loading the coupling circuit may introduce frequency-selective attenuation and disturb the whiteness of the noise spectrum. It is very important to adjust the transmitting coupler carefully to its loading impedance.

Narrowband interference features significant noise levels in the frequency domain in comparison with background noise. Envelopes of a number of narrowband interferers exhibit strong dynamic and time-varying features. A bandpass filtering based approach is proposed for the estimation of the time-varying envelopes. This approach can detect broadband impulsive noise and eliminate its influences on the estimation. The typical sinusoidal waveform of narrowband interference can be generated by using the phase accumulation method. The estimated envelopes can be modeled by unsymmetrical triangular functions. In addition to the typical narrowband interferer, a class of interferers with time-varying frequencies, called swept-frequency noise in this thesis, has been observed both in in-door and access domain. Such interference is usually caused by active power factor correction (PFC) circuits in power supply units of many end-user appliances, such as fluorescent lamps and PCs. Emulation is possible either by chirp functions or simulating PFC control signals and then filtering the synthesized mains current.

For impulsive noise, detection can be made by comparing the power of segmented noise waveforms with a constant threshold. In cases where the exact course of a noise waveform is unimportant, impulsive noise can be generated by switching the background noise on and off according to a control pulse sequence. Otherwise, it can be emulated as an exponentially attenuated oscillation. For periodic impulsive noise, the periodic switching sequence can be generated with a modulo-N counter. For aperiodic impulsive noise and burst noise, a partitioned Markov chain can be used to model the inter-arrival time, the pulse width and the noise level.

For colored background noise, an accurate estimation can only be made after all the other noise classes have been removed. Its power level depends on number and type of connected and active electrical devices. Colored background noise also exhibits cyclo-stationary characteristics synchronous with the mains frequency, and its smoothed spectrum can be approximated by the sum of two exponential functions.

7.5 Definition of Reference Channels

Typical and representative channel conditions have been collected from own measurements as well as from results reported in literatures. Four complex-valued CTFs are proposed as reference CTFs. With respect to noise scenarios, three levels, denoted by highly noisy, moderately noisy and quiet, are defined for each noise class. As the name suggests, the highly noisy scenario has the highest noise level, while the quiet scenario exhibits the lowest noise intensity.

7.6 Future Work

For future work it may be interesting to build extended PLC networks using multiple channel emulators. In this way, PLC system performance on the upper layers of communication can also be investigated in a convenient environment. Most of the proposed algorithms and approaches can also be applied to other frequency ranges, going up to e.g. 100 MHz, if needed.

Appendix A - Reference CTF

Unloaded LV Channels

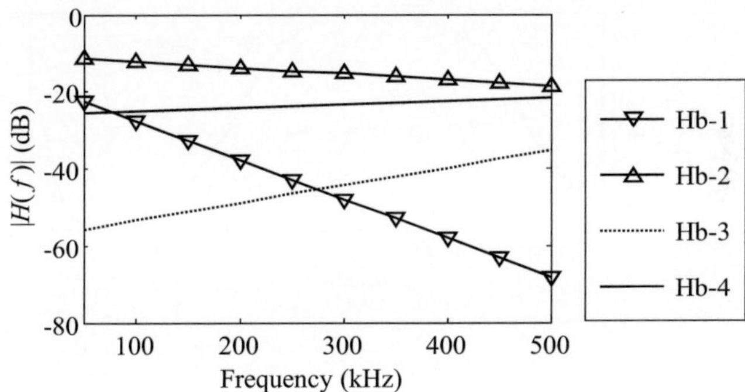

Fig. 7.1 Magnitude responses of unloaded channels. Hb-1: low pass, different branches, different phases [BAU05]; Hb-2: low pass, access domain, same phase; Hb-3: indoor channel, different branches, the same phase [SUG08]; Hb-4: indoor channel, same branch, different phases [SUG08].

Mathematical representation: A linear function used to describe the magnitude response of an unloaded LV power line channel.

$$| H \left(f \right) |_{dB} = H_0 \cdot f + H_1 \tag{7.1}$$

Table 7.1 Parameters of the Linear Function

Unloaded channel	$H_0[\cdot 10^{-6}]$	H_1	Description
Hb-1	-101.1	-17.6	low pass, different branches [BAU05]
Hb-2	-16.6	-10.4	low pass, same phase
Hb-3	45.5	-58	indoor channel, different branches [SUG08]
Hb-4	8.7	-25.8	indoor channel, same branch [SUG08]

Loaded LV channel: Hd-1

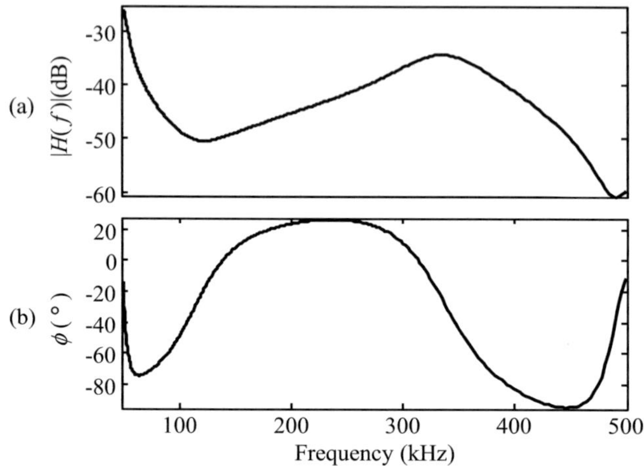

Fig. 7.2 Reference CTF Hd-1.

Table 7.2 Parameters for Cascaded Resonance Circuits

Stage	Circuit Type	R_0 (Ω)	R (Ω)	C (µF)	L (µH)
1	SRC	5	0.38	0.05	2.10
2	SRC	5	2.52	0.31	5.84
3	PRC	5	0.97	0.40	0.55
4	PRC	5	0.55	3.45	2.94

Loaded LV channel: Hd-2

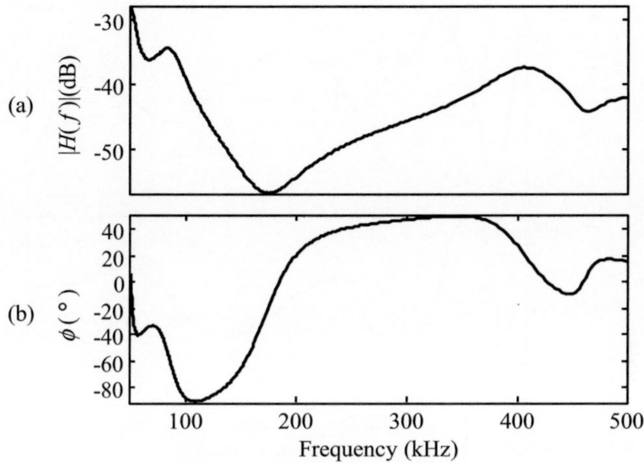

Fig. 7.3 Reference CTF Hd-2.

Table 7.3 Parameters for Cascaded Resonance Circuits

Stage	Circuit Type	$R_0\,(\Omega)$	$R\,(\Omega)$	$C\,(\mu F)$	$L\,(\mu H)$
1	SRC	5	13.31	$2.22 \cdot 10^{-3}$	54.07
2	PRC	5	3.13	0.33	0.46
3	PRC	5	1.56	1.27	2.64
4	PRC	5	1.80	3.35	2.90
5	SRC	5	14.63	$1.43 \cdot 10^{-3}$	83.25
6	SRC	5	1.74	0.15	5.51

Loaded LV channel: Hd-3

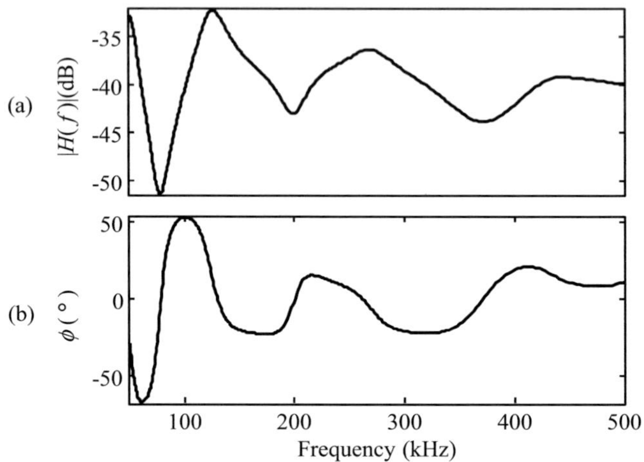

Fig. 7.4 Reference CTF Hd-3.

Table 7.4 Parameters for Cascaded Resonance Circuits

Stage	Circuit Type	R_0 (Ω)	R (Ω)	C (μF)	L (μH)
1	SRC	1	6.52	$5.06 \cdot 10^{-3}$	20.01
2	PRC	1	6.81	0.37	0.37
3	PRC	1	4.97	0.51	0.68
4	PRC	5	1.98	1.07	1.47
5	SRC	1	2.18	$4.70 \cdot 10^{-2}$	13.35
6	SRC	1	2.10	$3.79 \cdot 10^{-2}$	4.82
7	PRC	5	1.02	3.96	2.04
8	SRC	1	0.32	1.09	3.84

Loaded LV channel: Hd-4

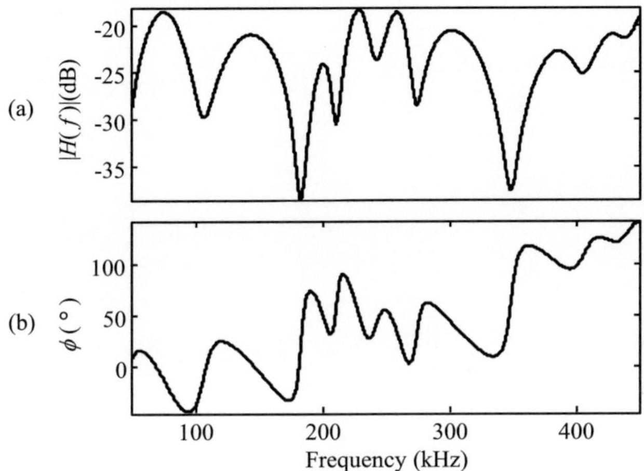

Fig. 7.5 Reference CTF Hd-4.

Table 7.5 Parameters for Cascaded Resonance Circuits

Stage	Circuit Type	$R_0\,(\Omega)$	$R\,(\Omega)$	$C\,(\mu F)$	$L\,(\mu H)$
1	SRC	2	1.93	$8.90 \cdot 10^{-3}$	14.63
2	SRC	2	1.21	$1.80 \cdot 10^{-2}$	8.54
3	SRC	2	0.17	$7.65 \cdot 10^{-2}$	2.73
4	SRC	1	0.39	0.57	3.93
5	SRC	2	0.25	2.64	4.73
6	SRC	1	1.49	$2.45 \cdot 10^{-2}$	17.73
7	SRC	1	5.52	$4.15 \cdot 10^{-3}$	135.60
8	SRC	1	1.35	$1.86 \cdot 10^{-2}$	30.79
9	SRC	1	1.11	$1.79 \cdot 10^{-2}$	18.93
10	SRC	1	4.22	$9.37 \cdot 10^{-3}$	81.50
11	SRC	2	0.59	$6.40 \cdot 10^{-2}$	11.80

Appendix B – Reference Noise Scenarios

Background noise

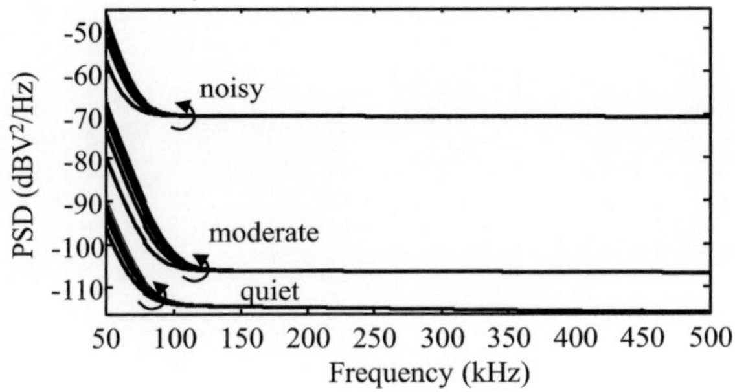

Fig. 7.6 PSDs of reference background noise.

The mathematical interpretation for the PSD of the colored background noise can be obtained according to (5.73) and (5.74)

$$\hat{P}_{BGN}(f) = \left[\frac{|m(t)|}{A_0} + A_1 \right] \cdot e^{b \cdot f} + c \cdot e^{d \cdot f} \tag{7.2}$$

Table 7.6 Parameters of Background Noise Model

Intensity	A_0	A_1	b	c	d
Noisy	$2 \cdot 10^2$	$4.41 \cdot 10^{-1}$	$-1.2 \cdot 10^{-1}$	$3.13 \cdot 10^{-4}$	$-1.32 \cdot 10^{-4}$
Moderate	$6.4 \cdot 10^3$	$1.5 \cdot 10^{-2}$	$-1 \cdot 10^{-1}$	$5 \cdot 10^{-6}$	$-2 \cdot 10^{-4}$
Quiet	$1.024 \cdot 10^5$	$2 \cdot 10^{-3}$	$-1 \cdot 10^{-1}$	$2 \cdot 10^{-6}$	$-5 \cdot 10^{-4}$

Narrowband noise (NBN)

Table 7.7 Parameters of the NBN Model

Intensity	f_m (kHz)	Δf (kHz)	A_{NBN} (dBμV)
Noisy	65	4	120
	80	6	110
	110	5	80
	125	11	70
	162.5	6	70
	207.5	8	70
	252.5	4	70
	297.5	5	70
	342.5	5	70
	432.5	5	70
	477.5	4	100
Moderate	65	4	90
	117.5	3	70
	162.5	2	60
	432.5	2	60
Quiet	75	3	70
	125	3	65

For the dynamic feature, the envelope of each NBN is

$$y(t) = A_{NBN} \cdot \left[y_1(t, t_{1,1}, t_{1,2}, t_{1,3}) + y_2(t, t_{2,1}, t_{2,2}, t_{2,3}) \right] \tag{7.3}$$

Table 7.8 Parameters of Dynamic NBN

f_m (kHz)	A_NBN (dBµV)	$t_{1,1}$(ms)	$t_{1,2}$(ms)	$t_{1,3}$(ms)	$t_{2,1}$(ms)	$t_{2,2}$(ms)	$t_{2,3}$(ms)
65	120	0.5	3	6	10.5	13	16
80	110	1	4	8	11	14	18
110	80	2	5	9	12	15	18
125	70	3	7	9	13	16	19
162.5	70	3	5	7	13	15	17

Swept-frequency noise

Table 7.9 Parameters of Swept-Frequency Noise

Intensity	$m_n(\mathrm{V})$	f_n(kHz)	f_0 (kHz)	f_1 (kHz)	τ_0 (ms)	$\Delta\tau$ (ms)
highly noisy	$\left. m_1(t)\right|_{\frac{T_M}{2}} = \dfrac{\lvert m(t)\rvert}{32500}$ $m_2(t) = 10^{-4}$	$f_0 + \dfrac{f_1 - f_0}{\Delta\tau}\cdot(\tau - \tau_0)$ $f_n(\tau) = f_n\!\left(\tau + k\cdot\dfrac{T_M}{2}\right)$	20	32	2	8
			32	20	2	8
			190	210	0	5
			210	190	5	5
	$m_3(t) = \dfrac{\lvert m(t)\rvert}{1625}$	$f_0 + f_1\cdot\left(\dfrac{\tau - \tau_0}{\tau_0}\right)^{1.8}$	45	145	2.2	-
moderate	$m(t) = 10^{-4}$	$f_0 + \dfrac{f_1 - f_0}{\Delta\tau}\cdot(\tau - \tau_0)$ $f_n(\tau) = f_n\!\left(\tau + k\cdot\dfrac{T_M}{2}\right)$	190	210	0	5
			210	190	5	5
quiet	-	-	-	-	-	-

Periodic impulsive noise

Table 7.10 Parameters of Periodic Impulsive Noise

Intensity	Repetition rate (Hz)	A_M (V)	T_F (µs)	f_M (kHz)	Delay to zero crossing (ms)
highly noisy	100	1.5	100	30	0
		1	100	40	1
		1	200	65	2
	20k	1	12	150	-
	12k	0.8	20	100	-
	40k	0.7	20	150	-
moderate	100	1	100	30	0
		0.5	100	40	1
	12k	0.5	20	100	-
	40k	0.5	20	150	-
quiet	100	0.5	100	30	0
	12k	0.5	20	100	-

Aperiodic impulsive noise

Table 7.11 Parameters for Reference APIN

Intensity	$\overline{t}_{D,1}$ (s)	$\overline{t}_{D,2}$ (s)	$\overline{t}_{D,3}$ (s)	$\overline{t}_{W,1}$ (µs)	$\overline{t}_{W,2}$ (µs)
highly noisy					
moderate	1	0.1	0.01	1000	100
quiet					

Table 7.12 Stationary Distribution for Reference APIN

Intensity	r_D %	$\pi_{u,1}$ %	$\pi_{u,2}$ %	$\pi_{u,3}$ %	$\pi_{g,1}$ %	$\pi_{g,2}$ %
highly noisy	1	1	8	90	0.1	0.9
moderate	0.1	1	90	8.9	0.01	0.09
quiet	0.01	90	8.99	1	0.001	0.009

Table 7.13 P-Matrix for Highly Noisy Scenario

State	1	2	3	4	5
1	16776377	0	0	9	830
2	0	16768827	0	92	8296
3	0	0	16693330	922	82964
4	**92**	**7390**	**831378**	15938355	0
5	**924**	**73900**	**8313784**	0	8388608

Table 7.14 P-Matrix for a Moderately Noisy Scenario

State	1	2	3	4	5
1	16776377	0	0	9	830
2	0	16768827	0	92	8296
3	0	0	16693330	922	82964
4	**468**	**421538**	**416854**	15938355	0
5	**4684**	**4215381**	**4168543**	0	8388608

Table 7.15 P-Matrix for a Quiet Scenario

State	1	2	3	4	5
1	16776377	0	0	9	830
2	0	16768827	0	92	8296
3	0	0	16693330	922	82964
4	**269730**	**269430**	**299700**	15938355	0
5	**2697302**	**2694305**	**2997002**	0	8388608

Burst noise

Table 7.16 Parameters for Reference Burst Noise

Intensity	t_{s_I} (ms)	$\bar{t}_{I,1}$ (ms)	$\bar{t}_{I,2}$ (ms)	$t_{S_{II}}$ (µs)	$\bar{t}_{II,1}$ (µs)	$\bar{t}_{I,2}$ (µs)
highly noisy			5			1000
moderate	1	1000	3	50	1000	500
quiet			1.01			100

Table 7.17 Transition Matrices for Reference Burst Noise

Intensity	P_I		P_{II}	
highly noisy	0.999	0.001	0.95	0.05
	0.2	**0.8**	**0.05**	**0.95**
moderate	0.999	0.001	0.95	0.05
	0.333	**0.667**	**0.1**	**0.9**
quiet	0.999	0.001	0.95	0.05
	0.99	**0.01**	**0.5**	**0.5**

Appendix C – Abbreviations

2PN	Two port network
AC	Alternating current
ADC	Analog-to-digital converter
AFE	Analog front-end
AGC	Automatic gain control
AM	Amplitude modulation
AMR	Advanced meter reading
APIN	Aperiodic impulsive noise
ARIB	Association of Radio Industries and Businesses
ARQ	Automatic repeat request
ASIC	Application-specific integrated circuit
AWGN	Additive white Gaussian noise
BB	Broadband
BER	Bit error rate
bps	bit per second
BPSK	Binary phase-shift keying
CC	Coupling circuits
CENELEC	Comité Européen de Normalisation Électrotechnique
CLT	Central limit theorem
CPFSK	Continuous-phase FSK
CSK	Code shift keying
CTF	Channel transfer function
DA	Distributed arithmetic

DAC	Digital-to-analog converters
DC	Direct current
DCSK	Differential code shift keying
DFT	Discrete Fourier transform
DIN	German institute for standardization (in German language: Deutsches Institut für Normung)
DLL	Data link layer
DPSK	Differential phase-shift keying
DSP	Digital signal processor
DUT	Devices under test
EMC	Electromagnetic compatibility
EMI	Electromagnetic interference
FCC	Federal Communications Commission
FCH	Frame control header
FER	Frame error rate
FFT	Fast Fourier transform
FIFO	First-in, first-out
FIR	Finite impulse response
FLR	Frame loss rate
FM	Frequency modulation
FPGA	Field programmable gate array
FSK	Frequency-shift keying
GPS	Global positioning system
HDD	Hard disk drive
IDFT	Inverse discrete Fourier transform
i.i.d	independent identically distributed
IEC	International electrotechnical commission

IP	Intellectual property
LFSR	Linear feedback shift registers
ISI	Inter-symbol interference
ISO	International organization for standardization
LISN	Line impedance stabilization networks
LTI	Linear time-invariant
LTV	Linear time-variant
LPTV	Linear periodically time-variant
LV	Low voltage
MAC	Media access control
MAC	Multiply and accumulate
MCU	Microcontroller unit
MIMO	Multiple-input and multiple-output
MLBS	Maximum length binary signals
MLS	Maximum length sequence
MSB	Most significant bit
MSK	Minimum-shift keying
MSps	Mega samples per second
MSV	Multi-cycle variable
MV	Medium voltage
NB	Narrowband
NL	Non-linear
OFDM	Orthogonal frequency division multiplexing
OPERA	Open PLC European Research Alliance
PAPR	Peak-to-average power ratio
PDF	Probability density function
PFC	Power factor correction

PGA	Programmable gain amplifier
PHY	Physical
PLC	Powerline communication
PLL	Phase-locked loop
PM	Phase modulation
PRC	Parallel resonant circuit
PRIME	Powerline intelligent metering evolution
PSD	Power spectral density
PSK	Phase-shift keying
PWM	Pulse-width modulation
QPSK	Quaternary phase-shift keying
RF	Radio frequency
RPC	Residential power circuit
SDO	Standard development organizations
SFN	Swept-frequency noise
SMPS	Switch mode power supply
SNR	Signal-to-noise ratio
SNRM	Signal to noise ratio margins
S-FSK	Spread frequency-shift keying
SoC	System on chip
SRC	Series resonant circuit
SPI	Serial peripheral interface bus
STFT	Short-time Fourier transform
UART	Universal asynchronous receiver/transmitter
UPS	Uninterruptible power supply
USB	Universal serial bus
VDSL	Very high speed digital subscriber line

VGA   Variable gain amplifier

VNA   Vector network analyzer

References

[ALT09]   *FIR compiler user guide*, ver. 9.1, Altera Co., San Jose, CA, 2009.

[ADD21]   *ADD1021 datasheet*, rev. 1.04, ADD Semiconductor, Zaragoza, Spain, May 2010.

[ADD22]   *ADD1022 brief datasheet*, rev. 1.02, ADD Semiconductor, Zaragoza, Spain, Apr. 2010.

[BAB05]   M. Babic, M. Hagenau, K. Dostert and J. Bausch, "Theoretical postulation of PLC channel models," Rep. D4 within the OPERA IST Integrated Project No. 507667, 2005.

[BAN02]   P. Banelli, G. Leus, and G. B. Giannakis, "Bayesian estimation of clipped Gaussian processes with application to OFDM," in *Proc. Eur. Signal Process. Conf.*, Toulouse, France, Sep. 2002, pp. 181–184.

[BAR04]   J. R. Barry, E. A. Lee and D.G. Messerschmitt, *Digital communication*, 3rd ed., Norwell, MA: Kluwer Academic Publishers, 2004.

[BAU05]   J. Bausch, *Elektrische Installationsnetze als Datenübertragungsmedium zur Gebäudeautomatisierung*, Berlin, Germany: Mensch & Buch Verlag, 2005.

[BAU052]   J. Bausch, T. Kistner, A. Moreau, S. Sauvage, and S. Milanini, "Advanced orphelec test equipment and novel test procedures," in *Proc. IEEE Int. Symp. Power Line Commun. Applicat.*, Udine, Italy, Apr., 2005, pp. 385-389.

[BAU06]   J. Bausch, T. Kistner, M. Babic, K. Dostert, "Characteristics of Indoor Power Line Channels in the Frequency Range 50-500 kHz," in *Proc. IEEE Int. Symp. Power Line Commun. Applicat.*, Orlando, FL, Mar. 2006.

[BOX58]   G. E. P. Box and M. E. Muller, "A Note on the Generation of Random Normal Deviates," in *Ann. Math. Statist.*, vol. 29, no. 2, 1958, pp. 610-611.

[CAN06]   F. J. Cañete, J. A. Cortés, L. Díez and J. T. Entrambasaguas, "Analysis of the Cyclic Short-Term Variation of Indoor Power Line Channels," *IEEE J. Sel. Areas Commun.*, vol. 24, no. 7, pp. 1327-1338, Jul. 2006.

[CAN08]   F. J. Cañete, L. Díez, J. A. Cortés, J. J. Sánchez-Martínez and L. M. Torres, "Time-Varying Channel Emulator for Indoor Power Line Communications," in *Proc. IEEE Global Commun. Conf.*, New Orleans, LA, 2008, pp. 2896-2900.

[CEN01]   *Signaling on low-voltage electrical installations in the frequency range 3 kHz to 148.5 kHz*, CENELEC EN 50065-1, 2002.

[CHA73]   D.S.K. Chan and L. R. Rabiner, "Analysis of quantization errors in the direct form for finite impulse response digital filters," *IEEE Trans. Audio Electroacoust.*, vol. AU-21, pp. 354-66, Aug., 1973.

[CHA732]   D.S.K. Chan and L. R. Rabiner, "Theory of roundoff noise in cascade realizations of finite impulse response digital filters," *Bell Syst. Tech. J.*, vol. 52, pp. 329-45, Mar. 1973.

[CHA86]   M. H. L.Chan and R. W. Donaldson, "Attenuation of communication signals on residential and commercial intrabuilding power-distribution circuits," *IEEE Trans. Electromagn. Compat.*, vol. 28, no. 4, pp. 220-230, Nov. 1986.

[CHO07]   W. Choi and C. Park, "A simple line coupler with adaptive impedance matching for power line communication," in *IEEE Int. Symp. Power Line Commun. Applicat.*, Pisa, Italy, 2007, pp. 187-191.

[CLA82]   T. A. C. M. Claasen and W. F. G. Mecklenbraeuker, "On stationary linear time-varying systems," *IEEE Trans. Circuits Syst.*, vol. 29, no. 3, pp. 169 - 184, Mar. 1982.

[COR10]   J.A. Cortes, L.Diez, F. J. Canete, and J.J. Sanchez-Martinez, "Analysis of the indoor broadband power-line noise scenario," *IEEE Trans. Electromagn. Compat.*, vol. 52, no. 4, pp. 849–858, Nov. 2010.

[CYP11]   *CY8CPLC20 powerline communication solution*, 001-48325 rev. *J, Cypress Semiconductor Corp., San Jose, CA, 2011.

[DEV86]   L. Devroye, *Non-uniform random variate generation*, New York: Springer-Verlag, 1986.

[COS09]   E. M. M. Costa, "A basic analysis about induced EMF of planar coils to ring coils," *Progress In Electromagn. Research B*, vol. 17, pp. 85-100, 2009.

[CTI11]   CTI, IBERDROLA, ITRON, L&G, ZIV, "PLC lower layer based on PRIME," Rep. WP5.3.1, ver. 1.0, deliverable of European project OPEN meter, Jun. 2011.

[DEI99]   M. Deinzer and M. Stöger, "Integrated PLC-modem based on OFDM," GmbH, Grosshabersdorf, in *Tech. Paper Int. Symp. Power Line Commun. Applicat.*, Lancaster, UK, Mar., 1999.

[DOS00]   K. Dostert, M. Zimmermann, T. Waldeck and M. Arzberger, "Fundamental properties of the low voltage power distribution grid used as a data channel," *European Trans. Telecommun.*, vol.11, May 2000, pp. 297-306.

[ECH01]   *Introduction to the LonWorks® platform*, rev. 2.0, Echelon Co., San Jose, CA.

[ECH02]   *LonWorks® PLT-22 transceiver user's guide (110 kHz-140 kHz operation)*, ver. 1.2, Echelon Co., San Jose, CA.

[ELL12]   R. Elliott. (2012, Oct. 06). *High voltage audio systems, internet Elliott Sound Products* [Online]. Available: http://sound.westhost.com/articles/line-amps.html#tst.

[FAI04]   *AN-42047, Power factor correction (PFC) basics, application note 42047*, rev. 0.9.0, Fairchild semiconductor, San Jose, CA, Aug. 2004.

[FAN04]   Y. Fan and Z. Zilic, "A novel scheme of implementing high speed AWGN communication channel emulators in FPGAs," in *Proc. Int. Symp. Circuits Syst.*, Rio de Janeiro, Brazil, May 2004, pp. 877-880.

[FAN08]   Y. Fan and Z. Zilic, "BER testing of communication interfaces," *IEEE Trans. on Instrum. Meas.*, vol. 57, no. 5, pp. 897-906, May 2008.

[FAZ03]   K. Fazel, and S. Kaiser, *Multi-carrier and spread spectrum systems: from OFDM and MC-CDMA to LTE and WiMAX*, 2nd ed., Chichester, UK: Wiley, 2008.

[FER10]   H. C. Ferreira, L. Lampe, J. Newbury and T. G. Swart, *Power line communications: theory and applications for narrowband and broadband communications over power lines.* Chichester, UK: Wiley, 2010.

[FUJ10]   *MB87S2080-PRIME-compliant PLC modem SoC*, Fujitsu limited, Tokyo, Japan, Sep. 2010.

[FUR02]   E.R. Furlong, "UPS topologies for large critical power systems (>500 KVA)," *in Proc. 13th Ann. Power Quality Exhibition & Conf.*, Oct. 2002.

[GAL01]   D. Galda, H. Rohling, "Narrow band interference reduction in OFDM-based power line communication systems," in *Tech. Paper Int. Symp. Power Line Commun. Applicat.*, Malmö, Sweden, 2001, pp. 345-351.

[GAL05]   S. Galli and T. Banwell, "A novel approach to accurate modeling of the indoor power line channel—Part II: transfer function and channel properties," *IEEE Trans. Power Del.*, vol. 20, no. 3, pp. 1869 - 1878, Jul. 2005.

[GAL06]   S. Galli and T. Banwell, "A deterministic frequency-domain model for the Indoor power line transfer function," *IEEE J.Sel. Areas Commun.*, vol. 24, no. 7, pp. 1304-1316, Jul. 2006.

[GAL11]   S. Galli, A. Scaglione and Z. Wang, "For the grid and through the grid: the role of power line communications in the smart grid," in *Proc. IEEE*, vol. 99, no. 6, pp. 998-1027, May, 2011.

[GAO09]   C. Gao and X. Cao, "Effects of group delay on the performance of OFDM system," *1st Int. Conf. Inform. Sci. Eng.*, Nanjing, China, Dec. 2009, pp. 2618 -2621.

[GOE04]   M. Götz, *Mikroelektronische, echtzeitfähige Emulation von Power-line-Kommunikationskanälen*, Berlin, Germany: Mensch & Buch Verlag, 2004.

[GRA01]   R. M. Gray and Lee D. Davisson, *An introduction to Statistical Signal Processing*, Cambridge, England: Cambridge University Press, 2004.

[HAE01]   E. Haensler, *Statistische Signale: Grundlage und Anwendungen*, 3rd ed., Berlin, Germany: Springer, 2001.

[HAR78]   F. J. Harris, "On the use of windows for harmonic analysis with the discrete Fourier transform," in *Proc. IEEE*, vol. 66, no. 1, pp. 51-83, Jan., 1978.

[HOC11]   M. Hoch, "Comparison of PLC G3 and PRIME," in *Proc. IEEE Int. Symp. Power Line Commun. Applicat.*, Udine, Italy, 2011, pp. 165-169.

[HOO98]   O.G. Hooijen, "A channel model for the residential power circuit used as a digital communications medium," *IEEE Trans. Electromagn. Compat.*, vol. 40, pp.331-336, Nov. 1998.

[IEC01]   *Distribution automation using distribution line carrier systems, part 5-1: Lower layer profiles. The spread frequency shift keying (S-FSK) profile*, IEC 61334-5-1:2001, 2001.

[JAC96]   Leland B. Jackson, *Digital Filters and Signal Processing*, 3rd ed., Morwell, MA: Kluwer Academic Publishers, 1996, pp. 289-307.

[JOE09]   K. Jones and C. Aslanidis, "DCSK technology vs. OFDM concepts for PLC smart metering," Renesas Electron. Co., Tokyo, Japan, Mar. 2009.

[KAI04]   K. L. Kaiser, *Electromagnetic compatibility handbook*, Boca Raton, FL: CRC Press, 2004.

[KAM09]   K. D. Kammeyer, K. Kroschel, *Digitale Signalverarbeitung*, 7th ed., Wiesbaden, Germany: Vieweg + Teubner, 2009.

[KAM11]   K. D. Kammeyer, *Nachrichtenübertragung*, 5th ed., Wiesbaden, Germany: Vieweg + Teubner, 2011.

[KAT06]   M. Katayama, T. Yamazato and H. Okada, "A Mathematical Model of Noise in Narrowband Power Line Communication Systems," *IEEE J. Sel. Areas Commun.*, vol. 24, no. 7, pp. 1267-1276, Jul. 2006.

[KER87]   R.J. Kerczewski, "A study of the effect of group delay distortion on an SMSK satellite communications channel," NASA, Washington, D.C., Rep. 89835, Apr. 1987.

[KIS08]   T. Kistner, M. Bauer, A. Hetzer, K. Dostert, "Analysis of zero crossing synchronization for OFDM-based AMR systems," in *Proc. IEEE Int. Symp. Power Line Commun. Applicat.*, Jeju Island, Korea, Apr., 2008, pp. 204-208.

[LAR11A]   A. Larsson, "On high-frequency distortion in low-voltage power systems," doctoral thesis, Universitetstryckeriet, Luleå, Sweden, 2011.

[LAR11B]   A. Larsson, M. Bollen, "Emission (2 to 150 kHz) from a light installation," in *Proc. 21st Int. Conf. on electricity distribution*, Frankfurt, Germany, Jun., 2011.

[MAG88]   *Soft Ferrites, A Users Guide*, Magnetic Materials Producers Association, 1998.

[MAR64]   G. Marsaglia, T. A. Bray, "A convenient method for generating normal variables," in *Soc. Ind. Appl. Math. Review*, vol.6, no.3, Jul., 1964.

[MAT11]   P. Mathews. (2011, Feb. 02). *High voltage audio: unwinding distribution transformers* [Online]. Available: http://www.prosoundweb.com-/article/high_voltage_audio/P2/.

[ENE11]   ENEL, "PLC lower layers on B-PSK for the Meters and More suite," Rep. WP5.3.3, ver. 1.0, deliverable of European project OPEN meter, Jun. 2011.

[ERD11]   ERDF, EDF R&D, "PLC lower layer based on G3," Rep. WP5.3.2, ver. 1.0, deliverable of European project OPEN meter, Jun. 2011.

[PHI99]   H. Philipps, "Modeling of Powerline Communication Channels," in *Tech. Paper Int. Symp. Power Line Commun. Applicat.*, Lancaster, UK, Mar., 1999. pp. 14-21.

[PRO01]   J. G. Proakis, *Digital communications*, 4th ed., Boston, MA: McGraw Hill, 2001.

[RAB75]   L.R. Rabiner and B. Gold, *Theory and application of digital signal processing*, Englewood Cliffs, NJ: Prentice-Hall, 1975.

[RIC95]   J.Rice, *Mathematical statistics and data analysis,* 2nd ed., Belmont, CA: Duxbury Press, 1995.

[SAN07]   S. Sancha, F.J. Canete, L. Diez, and J.T. Entrambasaguas, "A channel simulator for indoor power-line communications," in *Proc. IEEE Int. Symp. Power Line Commun. Applicat.*, Pisa, Italy, Mar. 2007, pp. 104-109.

[SMI99]   S.W. Smith, *The scientist and engineer's guide to digital signal processing*, 2nd ed., San Diego, CA: California Technical Publishing, 1999.

[SMI11]   J. O. Smith. (2012, Jan. 15). *Spectral Audio Signal Processing* [Online]. Available: http://ccrma.stanford.edu/~jos/sasp/.

[ST7538]   *FSK power line transceiver, ST7538 data sheet*, 5th ed., STMicroelectronics N. V., Geneva, Switzerland, Nov. 2005.

[ST7540]   *FSK power line transceiver, ST7540 data sheet*, 2nd ed., STMicroelectronics N. V., Geneva, Switzerland, Sep. 2006.

[ST7580]   *FSK, PSK multi-mode power line networking system-on-chip, ST7580 data sheet*, 1st ed., STMicroelectronics N. V., Geneva, Switzerland, Jan. 2012.

[ST7590]   *Narrowband OFDM power line networking PRIME compliant system-on-chip, ST7590 data sheet*, 1st ed., STMicroelectronics N. V., Geneva, Switzerland, Oct. 2011.

[SUG08]   Y. Sugiura, T. Yamazato and M. Katayama, "Measurement of narrowband channel characteristics in single-phase three-wire indoor power-line channels," in *Proc. IEEE Int. Symp. Power Line Commun. Applicat.*, Jeju Island, Korea, Apr. 2008, pp. 18-23.

[SUG09] Y. Sugiura, T. Yamazato and M. Katayama, "Phase compensation for narrowband PLC channels with cyclic time-varying features," in *Proc. 3rd Workshop Power Line Commun.*, Udine, Italy, Oct. 2009.

[SUT98] P. Sutterlin, "A Power Line Communication Tutorial: Challenges and technologies," in *Tech. Paper Int. Symp. Power Line Commun. Applicat.*, Tokyo, Japan, Mar. 1998, pp. 24-26.

[TEX11] *TI PLC development kit user guide*, rev. 0.9, Texas Instruments, Dallas, TX. Sep. 2011.

[THO07] D. B. Thomas, W. Luk, P. H. Leong, J. D. Villasenor, "Gaussian random number generators," *Assoc. Comput. Mach. Comput. Surveys*, vol. 39, no. 4, 2007.

[TIE09] U. Tietze, C. Schenk, and E. Gamm, *Halbleiter-Schaltungstechnik*, 13th ed., Berlin; Heidelberg, Germany: Springer, 2009.

[VAN90] D. Van Wassenhove and J. Schoukens, "Group delay measurements of a transmission line using single channel measurements," *IEEE Trans. Instrum. Meas.*, vol. 39, pp.180-183, 1990.

[WEL11] N. Weling, "Flexible FPGA-Based Powerline Channel Emulator for Testing MIMO-PLC, Neighborhood Networks, Hidden Node or VDSL Coexistence Scenarios," in *Proc. IEEE Int. Symp. Power Line Commun. Applicat.*, Udine, Italy, Apr. 2011, pp. 12-17.

[WHE28] H. A. Wheeler, "Simple inductance formulas for radio coils", Proc. Inst. Radio Eng., vol. 16, no. 10, Oct. 1928.

[WOO05] G. Woods and M. Douglas, "Improving group delay measurement accuracy using the FM envelope delay technique," in *Proc. IEEE Region 10 Conf.*, Melbourne, VIC, Australia, Nov. 2005, pp. 1-6.

[XIL11] *LogiCORE IP fast Fourier transform, product specification*, ver. 7.1, XILINX, San Jose, CA, Mar. 2011.

[YU09] C. Yu and Y. Chen, "A new measurement method for group delay of frequency-translating devices based on software defined radio technology," in *Proc. 9th Int. Conf. Electron. Meas. Instrum.*, Beijing, China, Aug. 2009.

[ZIM02]   M. Zimmermann and K. Dostert, "Analysis and modeling of impulsive noise in broad-band powerline communications," *IEEE Trans. Electromagn. Compat.*, vol. 44, no. 1, pp. 249-258, 2002.

List of Publications

[BAU09]   M. Bauer, **W. Liu** and K. Dostert, "Channel emulation of low-speed PLC transmission channels," in *Proc. IEEE Int. Symp. Power Line Commun. Applicat.*, Dresden, Germany, Mar. 2009, pp. 267-272.

[BAU10]   M. Bauer, and **W. Liu**, "Robustness testing of an FSK PLC physical layer implementation by means of a channel emulator," In *Fernando Puente and Klaus Dostert editors, Rep. on Ind. Inform. Tech.*, Karlsruhe, Germany: KIT Scientific Publishing, 2010, vol. 12, pp. 113-128.

[KIS08]   T. Kistner, M. Bauer, J. Kallenberg, T. Fath and **W. Liu**, " Design of a PLC Modem for Automation of Mining Systems," In *Uwe Kiencke and Klaus Dostert editors, Berichte aus der Informatik Rep. on Ind. Inform. Tech.*, Aachen, Germany: Shaker Verlag, 2008, vol. 11, pp. 91-102.

[KIS082]   T. Kistner and **W. Liu**, "Error Detection and Error Location at Damaged Transmission Lines," In *Uwe Kiencke and Klaus Dostert editors, Berichte aus der Informatik Rep. on Ind. Inform. Tech.*, Aachen, Germany: Shaker Verlag, 2008, vol. 11, pp. 103-116.

[LIU10]   **W. Liu** and Y. Bao, "Least-squares-estimation based zero-crossing prediction for an automated PLC channel measurement system," In *Fernando Puente and Klaus Dostert editors, Rep. on Ind. Inform. Tech.*, Karlsruhe, Germany: KIT Scientific Publishing, 2010, vol. 12, pp. 129-145.

[LIU102]   **W. Liu** and K. Dostert, "Performance testing of PLC systems by using a PLC channel emulator," in *Proc. 4th Workshop Power Line Commun.*, Boppard, Germany, Sep. 2010.

[LIU11]   **W. Liu**, M. Sigle and K. Dostert, "Advanced Emulation of Channel Transfer Functions for Performance Evaluation of Powerline Modems," in *Proc. IEEE Int. Symp. Power Line Commun. Applicat.*, Udine, Italy, Apr. 2011, pp. 446-451.

[LIU111]   **W. Liu**, C. Li and K. Dostert, "Emulation of AWGN for noise margin test of powerline communication systems," in *Proc. IEEE Int. Symp. Power Line Commun. Applicat.*, Udine, Italy, Apr. 2011, pp. 225-230.

[LIU113]   **W. Liu** and K. Dostert, "Development and application of GPS-based time synchronization in power line channel characterization," in *Proc. 5th Workshop Power Line Commun.*, Arnhem, The Netherlands, Sep. 2011.

[LIU114]   **W. Liu**, M. Sigle and K. Dostert, "Channel characterization and system verification for narrowband power line communication in smart grid applications," *IEEE Commun. Mag.*, vol. 49, pp. 28-35, Dec. 2011.

[LIU12]   **W. Liu**, M. Sigle and K. Dostert, "Channel phase distortion and its influence on PLC systems," in *Proc. IEEE Int. Symp. Power Line Commun. Applicat.*, Beijing, China, Mar. 2012, pp. 268-273.

[SIG11]   M. Sigle, M. Bauer, **W. Liu** and K. Dostert, "Transmission Channel Properties of the Low Voltage Grid for Narrowband Power Line Communication," in *Proc. IEEE Int. Symp. Power Line Commun. Applicat.*, Udine, Italy, Apr. 2011, pp. 289-294.

[SIG12]   M. Sigle, **W. Liu** and K. Dostert, "On the impedance of the low-voltage distribution grid at frequencies up to 500 kHz," in *Proc. IEEE Int. Symp. Power Line Commun. Applicat.*, Beijing, China, Mar. 2012, pp. 30-34.

List of Supervised Student Assignments

[BAO09]   Y. Bao, "Design and implementation of an automatic measurement system for PLC," graduate thesis, Inst. of Ind. Inform. Tech. (IIIT), Karlsruhe Inst. of Tech.(KIT), 2009.

[FIG09]   P. Figuli, " Entwicklung eines DSP-Systems zur digitalen Audiosignalverarbeitung," Studienarbeit, Inst. of Ind. Inform. Tech. (IIIT), Karlsruhe Inst. of Tech.(KIT), 2009.

[LI09]   C. Li, "Design and verification of a FPGA-based noise generator," graduate thesis, Inst. of Ind. Inform. Tech. (IIIT), Karlsruhe Inst. of Tech.(KIT), 2009.

[WAN09]   J. Wang, "Amplitude and phase control for an automatic PLC measurement system," graduate thesis, Inst. of Ind. Inform. Tech. (IIIT), Karlsruhe Inst. of Tech.(KIT), 2009.

[DAR10]   D. Delic, "Drehzahlmessung mit dem Mikrocontroller ADuC 7026 für das Praktikum ‚Mikrocontroller und digitale Signalprozessoren'," Studienarbeit, Inst. of Ind. Inform. Tech. (IIIT), Karlsruhe Inst. of Tech.(KIT), 2010.

[SHA10]   K. Shakerian, "Digitale Frequenzsynthese mit dem ADuC7026," Studienarbeit, Inst. of Ind. Inform. Tech. (IIIT), Karlsruhe Inst. of Tech.(KIT), 2010.

[WAN10]   B. Wang, " Realisierung einer bidirektionalen Kommunikation für eine PLC-Testumgebung," Studienarbeit, Inst. of Ind. Inform. Tech. (IIIT), Karlsruhe Inst. of Tech.(KIT), 2010.

[XU10]   M. Xu, "Implementation of a high-speed data acquisition module for a PLC channel measurement system," graduate thesis, Inst. of Ind. Inform. Tech. (IIIT), Karlsruhe Inst. of Tech.(KIT), 2010.

[YAN10]   P. Yang, "Evaluation of mains voltage as a synchronization reference for PLC-AMS," graduate thesis, Inst. of Ind. Inform. Tech. (IIIT), Karlsruhe Inst. of Tech.(KIT), 2010.

[GAR11]   E. V. Garcia, "Development of a timekeeping algorithm for a GPS-based synchronization concept," student project, Inst. of Ind. Inform. Tech. (IIIT), Karlsruhe Inst. of Tech.(KIT), 2011.

**Forschungsberichte
aus der Industriellen Informationstechnik
(ISSN 2190-6629)**

Institut für Industrielle Informationstechnik
Karlsruher Institut für Technologie

Hrsg.: Prof. Dr.-Ing. Fernando Puente León, Prof. Dr.-Ing. habil. Klaus Dostert

Die Bände sind unter www.ksp.kit.edu als PDF frei verfügbar oder als Druckausgabe bestellbar.

Band 1 Pérez Grassi, Ana
Variable illumination and invariant features for detecting and classifying varnish defects. (2010)
ISBN 978-3-86644-537-6

Band 2 Christ, Konrad
Kalibrierung von Magnet-Injektoren für Benzin-Direkteinspritzsysteme mittels Körperschall. (2011)
ISBN 978-3-86644-718-9

Band 3 Sandmair, Andreas
Konzepte zur Trennung von Sprachsignalen in unterbestimmten Szenarien. (2011)
ISBN 978-3-86644-744-8

Band 4 Bauer, Michael
Vergleich von Mehrträger-Übertragungsverfahren und Entwurfskriterien für neuartige Powerline-Kommunikationssysteme zur Realisierung von Smart Grids. (2012)
ISBN 978-3-86644-779-0

Band 5 Kruse, Marco
Mehrobjekt-Zustandsschätzung mit verteilten Sensorträgern am Beispiel der Umfeldwahrnehmung im Straßenverkehr (2013)
ISBN 978-3-86644-982-4

Band 6 Dudeck, Sven
Kamerabasierte In-situ-Überwachung gepulster Laserschweißprozesse (2013)
ISBN 978-3-7315-0019-3

Band 7 Liu, Wenqing
Emulation of Narrowband Powerline Data Transmission Channels and Evaluation of PLC Systems (2013)
ISBN 978-3-7315-0071-1